Power Despite Precarity

"A masterful look at the challenges involved with organizing workers in higher education. Berry and Worthen provide excellent recommendations regarding vision and strategy, making the book valuable beyond the field of higher education."
—Bill Fletcher, Jr., author of *They're Bankrupting Us:
And Twenty Other Myths About Unions*

"Academic precarity screws over college and university teachers, partly via obfuscation about whether better working conditions are possible and partly by stealing our access to institutional memories of how precarious workers have risen up to win those better conditions—or something closer to them—in the past. Who fought for something better? How did they define what 'better' meant? What strategy and tactics did they use to make progress? What didn't work, wasn't worth fighting for, or wasted time along the way? *Power Despite Precarity* is an essential primer on these questions and more; a must-read for new adjuncts and organizers, as well as for movement veterans seeking a clear and coherent telling of the story of which they are a part, and some sparing but wise words of advice along the way."
—Alyssa Picard, Director, American Federation of
Teachers' higher education division

"Empowers us to fight for the higher education and unions we believe in, uniting theory and practice to chart an inspiring path toward labor and education justice."
—Mia L. McIver, Ph.D., Lecturer, UCLA, President,
University Council-American Federation of Teachers

"Written from both an organizer's and historian's perspective, *Power Despite Precarity* is essential reading for anyone working in higher education who wants a better world and wonders what it takes. Berry and Worthen provide a handbook on how the growing number of contingent faculty can unite in common cause. While it is about education, many of the lessons dealing with internal problems inside unions are not issues confined to the education sector (alas) and I especially enjoyed those parts."
—Elaine Bernard, Fellow, Labor & Worklife Program, Harvard Law School

"This is not just an important book but an essential one for anyone concerned about higher education. It is impossible to separate the working conditions of faculty from the learning conditions of students, and Berry and Worthen explain how it is possible to transform both for the better of all."
—Maria Maisto, President, New Faculty Majority

Wildcat: Workers' Movements and Global Capitalism

Series Editors:
Immanuel Ness (City University of New York)
Peter Cole (Western Illinois University)
Raquel Varela (Instituto de História Contemporânea [IHC]
of Universidade Nova de Lisboa, Lisbon New University)
Tim Pringle (SOAS, University of London)

Power Despite Precarity

Strategies for the Contingent Faculty Movement in Higher Education

Joe Berry and Helena Worthen

First published 2021 by Pluto Press
345 Archway Road, London N6 5AA

www.plutobooks.com

Copyright © Joe Berry and Helena Worthen 2021

The right of Joe Berry and Helena Worthen to be identified as the authors of this work has been asserted in accordance with the Copyright, Designs and Patents Act 1988.

British Library Cataloguing in Publication Data
A catalogue record for this book is available from the British Library

ISBN	978 0 7453 4553 6	Hardback
ISBN	978 0 7453 4552 9	Paperback
ISBN	978 0 7453 4556 7	PDF
ISBN	978 0 7453 4554 3	EPUB
ISBN	978 0 7453 4555 0	Kindle

Typeset by Stanford DTP Services, Northampton, England

Simultaneously printed in the United Kingdom and United States of America

Contents

PART V   SEVEN TROUBLESOME QUESTIONS

PART VI   USING THE POWER WE HAVE

Photographs

(pages 149–154)

Series Preface

Workers' movements are a common and recurring feature in contemporary capitalism. The same militancy that inspired the mass labor movements of the twentieth century continues to define worker struggles that proliferate throughout the world today.

For more than a century, labor unions have mobilized to represent the political-economic interests of workers by uncovering the abuses of capitalism, establishing wage standards, improving oppressive working conditions, and bargaining with employers and the state. Since the 1970s, organized labor has declined in size and influence as the global power and influence of capital has expanded dramatically. The world over, existing unions are in a condition of fracture and turbulence in response to neoliberalism, financialization, and the reappearance of rapacious forms of imperialism. New and modernized unions are adapting to conditions and creating class-conscious workers' movement rooted in militancy and solidarity. Ironically, while the power of organized labor contracts, working-class militancy and resistance persists and is growing in the Global South.

Wildcat publishes ambitious and innovative works on the history and political economy of workers' movements and is a forum for debate on pivotal movements and labor struggles. The series applies a broad definition of the labor movement to include workers in and out of unions, and seeks works that examine proletarianization and class formation; mass production; gender, affective and reproductive labor; imperialism and workers; syndicalism and independent unions, and labor and Leftist social and political movements.

Acknowledgements

Both of us would like to begin by thanking our parents. On both sides, our mothers and our fathers were teachers. Their lives would have been very different if they had had a union.

Then we thank John Hess, in memory, his son Andy Hess and Gail Sullivan's son Sean Sullivan. We thank Catherine Powell, the Director of the San Francisco State University Labor Archives.

Un-named individually but essential to this project are the activists who are engaged in this struggle across North America, including the now 3,000 plus subscribers to COCAL Updates who have read and forwarded this news aggregator for over 20 years now. They have also written reports, sought and given advice and generally been the backbone of this movement as well as serving as guinea pigs for innumerable academic survey studies about contingent faculty.

This book is a channel of movement knowledge. Although we are responsible to a great extent for what is included, we are not going to say that all mistakes are our responsibility. They are not. Some of the errors and misapprehensions that will be cringe-worthy ten years from now are not our fault. This is one of the consequences of something being a collective project. Fixing those errors in print or in practice will be the responsibility of whoever writes the next book and makes new errors. We hope this extended acknowledgement section demonstrates a bit of how movement or union learning happens, with a bow to D'Arcy Martin for coining the term, "union learning." This concept is discussed at much greater length in Helena Worthen's book *What Did You Learn at Work Today*.

In many ways this book is a sequel to Joe's 2005 book, *Reclaiming the Ivory Tower (RIT)*, which was about organizing contingent faculty. It was published the year after the 2004 COCAL VI conference in Chicago where we first tried to define a strategy for the contingent faculty movement. Since the publication of *RIT*, people have asked for something that talks about the work of representing contingent faculty and the movement overall. A lot of the same people who helped us with that book and then pushed us to do this one are still around, although older. So to start we want to thank again all those who agreed to be interviewed for RIT, especially Earl Silbar from AFSCME Adult Educators in Chicago City Colleges and Tom Suhrbur from IEA/NEA. Steve Hiatt and Beverly Stewart were Joe's key readers and editors before that book was handed off to Monthly Review Press.

Therefore we want to start with thanking our friends in Chicago who have stayed in the fight and contributed to the book which you are reading now, more than 15 years later. We include among these Beverly Stewart, Frank Rogaczewski, Frank Brooks, Joe Fedorko and LuAnn Schwartzlander at RAFO; Jocelyn Graf; Richard Schneirov in Indiana; Curtis Keyes at East-West University, Diane Stokes and Steve Edwards at AFSCME 2858; Rick Packard, Marie Cassady, and John Bolter at CCCLOC; Tony Johnson at AFT 1600; Tom Gradel, Karen Ford and Sergio Finardi of the NWU; and Penny Pixler of the IWW. In Champaign: our colleagues Martha Glotzhober, Toby Higbee, Gene Vanderport, Jim and Jenny Barrett, and especially Ed Hertenstein, our Director of Labor Ed at the School for Labor and Employment Relations, and Joel Cutcher-Gerschenfield, who was Dean while we were there. Like a lot of people, we took inspiration from the Chicago Teachers Union, CTU/AFT Local 1, their 2012 and subsequent strikes.

The following people have played multiple roles in our lives over many years. Among them are Fred Glass, who has been at various times our colleague, editor, supervisor, fellow labor educator, friend and comrade. Steve Hiatt, our long time friend, read the entire manuscript several times and has been an ongoing strategic advisor about everything related to publishing. Rodger Scott, colleague of many decades, former AFT 2121 union president, author and documentary film maker, persistently encouraged Joe Berry to write more and get a PhD. Maria Teresa Lechuga, who has taught at the UNAM (Autonomous University of Mexico) for 20 years in the Acatlán Faculty of Higher Studies. and Arturo Ramos Almazan, who is a professor at UNAM and was President of the University of Chapingo Academic Union (2001–2003), are friends, colleagues, and have provided essential support for the participation of Mexican faculty in the COCAL International conferences. Brian Simmons, carpenter and contingent faculty member for many years before finally getting hired full-time, helped us keep bikes, phones and computers running and asked the best questions. Frank Cosco, former president of the Vancouver Community College Faculty Association, where contingents have the best contract in North America, inspired this book on the best contract in the United States. He also co-authored the *Program for Change* with Jack Longmate, to whom we also owe deep thanks. Jonathan Karpf read this manuscript, corrected many errors and persuaded the CFA to subsidize this project financially and politically. Gary Rhoades of the University of Arizona has published some of the best research on faculty and their unions and has encouraged us in our work for many years.

There are also groups, networks and organizations that have been a home base, sometimes a refuge for us and others in this movement. These are places where the goal of the elimination of contingency is not debatable. The COCAL International Committee is an evolving group we have been part of since 1999, and without them this book definitely would not exist. Of that group especially Marcia Newfield, Vinnie Tirelli and David Rives have been consistent supporters. Emerging from COCAL, the Campus Equity Week national committees and the workers on those committees all over America since 2000, especially in California and Illinois, have been important. We have been long-time members of the California Federation of Teachers, especially the Part-timers Committee (where Joe and Helena met in the late 1980s) and the Coalition to Save City College. For Joe, AFT 2121 and recently the Retirees' Chapter have been critically important. Both former President Alisa Messer and now-retired Executive Director Chris Hanzo have discussed many of the ideas in this book over the years. The Labor Notes network, its publications and conferences, has been essential for us, as it has for thousands of others. The New Faculty Majority National Board, including Judy Olson, Anne Wiegard, and Robin Sowards, led by Maria Maisto, has provided an example of how to apply our strategies in real life. The Service Employees International Union's Adjunct Action AKA Faculty Forward campaign, led nationally by Melini Cadambi Daniels, is patterned after the pioneering work done at SEIU Local 500 in Washington, DC, and Anne McLeer. The United Association for Labor Education and its journal, *Labor Studies Journal*, has enabled us to publish and find colleagues in the United States and Canada, in particular long-time friend Bruce Nissen, faculty activist in Florida. The *Journal of Labor and Society*, formerly *WorkingUSA*, edited by Manny Ness, has published our work on this and other labor topics and appointed us both to their editorial board. We thank him for inviting us to submit our manuscript to Pluto's Wildcat series. We have also benefited from the behind-the-scenes, priceless work of the San

Francisco Bay Area Labor History Workshop and the Illinois Labor History Society. Last but not least, Joe thanks the old labor guys in the Bay Area BOYZ Group.

There are also important places where ideas in this book have been debated, developed and clarified, where participants have a specific stake in these issues but where our view is not taken for granted. The work here ranges from common projects, movement campaigns, to writing and publishing articles. These include the San Francisco Bay Area Higher Educators United; the National AAUP and its Committee on Contingency and the Professions and Joint Committee on Contingent Faculty and Governance, chaired by Mayra Besosa of CFA; the Organization of American Historians Committee on Part-time, Adjunct and Contingent Employment; the California Part Time Faculty Association and its conferences; the Faculty Association of California Community Colleges conferences; the Campaign for the Future of Higher Education; the Left Forum conferences and journal, *Radical Teacher*; the National Center for the Study of Collective Bargaining in Higher Education, especially Michelle Savarese, co-author with Joe Berry of the 2012 directory of faculty unions and contracts; the Coalition on the Academic Workforce; the Part-time and Contingent Faculty Caucus of National AFT and Alyssa Picard of AFT; the Center for the Study of Academic Labor, Colorado St. U, and the Contingent Faculty Committee of the NEA.

Of course, the primary organization that informed this project was the California Faculty Association. John Hess was able to provide both the personal experience and the contacts in CFA who could fill in a retrospective participant action research case study. In addition to the people who were interviewed and quoted at length in this book, we want to thank Alice Sunshine, Lil Taiz, Craig Flannery, Chris Cox and Vincent Cevasco who provided some of the photos and IDs.

We especially want to thank the people who agreed to be interviewed: Susan Meisenhelder, Nina Fendel, Jane Kerlinger, Jack Kurzweil, Katie Quan, Jonathan Karpf, Anne Robertson, Elizabeth Hoffman, Gretchen Reavy, Craig Flannery, and Bob Muscat.

People who read and commented on early drafts include Marcy Rein, Michael Mauer, Gary Zabel, James Tracy, Steven Herzenberg, Richard Moser, Robert Ovetz, David Kellogg, Fang Li and Earl Silbar.

Many people listened critically to our thumbnail summaries, gave us places to stay, interviewed us and asked hard questions, loaned us a car or otherwise provided material support. These include Kathy Kahn, Gail Sullivan, Mona Field, Martin Goldstein, Leanna Noble and Hollis Stewart, Pamela Vossenas and Michael Italie, Frank Bardacke and Julie Miller, Gary Zabel, Harry Brill, Heather Reimer, Jonathan Kissam, Bill Shields, Toni Mester, Mark Greenside, Bob Gabriner, Michael Zweig, Carol Stein, Cita Cook, David Slavin, Barbara Wolf, Ishmael Minune, Wendy Rader-Konofalski, Barbara Byrd, David Rives, Bob Bezemek, Keith Hoeller, Gifford Hartman, John Blanchette, Don Ehron, Sue Doe, Suzanne Hudson and Patrice Lawless.

Finally, the team at Pluto has been incredible. After a COVID-delayed start, our experience working with Pluto has been marked by seriousness, promptness, good communication and the sort of professionalism that comes from seeing a project as part of a movement, not a commercial venture.

Joe Berry and Helena Worthen
Berkeley, California
May 2021

Abbreviations and Acronyms

AAPAUNAM	*Asociación Autonóma del Personal Académico de la Universidad Autonoma de México. Syndicato de Insitucion de Universidad Autonóma de México*
AAUP	American Association of University Professors
ABD	All But Dissertation (completed course work, qualifying exams, for PhD, etc.)
ACCJC	Accrediting Commission for Community and Junior Colleges
ACORN	Association of Community Organizations for Reform Now
AFL-CIO	American Federation of Labor and Congress of Industrial Organizations
AFSCME	American Federation of State, County and Municipal Employees
AFT	American Federation of Teachers, part of the AFL-CIO
CAPI	California Association of Part-Time Instructors
CAW	Coalition on the Academic Workforce, a coalition formed by national professional disciplinary organizations, later joined by national unions and including the NFM to do research and publicity around the transformation of the academic workforce into contingent status. Publisher of major study in 2012.
CCCLOC	City Colleges Contingent Labor Organizing Committee (of Chicago, IEA/NEA)
CCSF	City College of San Francisco
CFA	Congress of Faculty Associations, the name of the proposed bargaining agent in the CSU election of 1982 which was a coalition of the local affiliates NEA/CTA, AAUP and CSEA. The name was changed after the election to CFA (same initials) meaning the California Faculty Association.
CFHE	Coalition on the Future of Higher Education

CFT	California Federation of Teachers, affiliated with the AFT/ AFL-CIO; United Professors of California (UPC) was the local CFT/AFT affiliate in the CSU system.
CGEU	Coalition of Graduate Employee Unions (as of 2021 changing its name to Coalition of Student Employee Unions)
CIO	Congress of Industrial Organizations
CNA	California Nurses Association, now part of National Nurses United, NNU/AFL-CIO
COBRA	Consolidated Budget Reconciliation Act
COCAL	Coalition of Contingent Academic Labor
CORE	Caucus of Rank-and-File Educators
CPFA	California Part-time Faculty Association
CSEA	California State Employees Association, affiliated since 1983 with SEIU as Local 1000
CSU	California State University (system), referred to as "the CSUs"
CTU	Chicago Teachers Union
CUNY	City University of New York
DA	Delegate Assembly, the representative body in many unions including the CFA
DRUM	Dodge Revolutionary Union Movement in Detroit, 1968–71. Later became part of the Black Workers Congress.
EDD	Employment Development Department
EERA	Education Employment Relations Act of California, one of the California public employment relations laws, 1976, giving legal collective bargaining rights to K–12 and Community College District employees. Referred to as the "Rhodda Act."
FICA	Federal Insurance Contributions Act. The abbreviation that appears on employee paychecks to indicate the deduction for Social Security and in some places Medicare.
FLSA	Fair Labor Standards Act, 1938
FMLA	Family Medical Leave Act, 1993
FSM	Free Speech Movement
HEERA	Higher Education Employer-Employee Relations Act, 1979. The California collective bargaining law that gave collective

	bargaining rights to higher education public employees in the CSUs and the University of California system.
IBT	International Brotherhood of Teamsters, sometimes affiliated with AFL-CIO
IWW	Industrial Workers of the World
I/Os	Inside/Outside strategy
MA	Master of Arts
MES	More Effective Schools
MFD	Miners for Democracy, reform movement within UMWA
MFDP	Mississippi Freedom Democratic Party
MLA	Modern Language Association
NAFFE	North American Alliance for Fair Employment
NAFTA	North American Free Trade Agreement
NEA	National Education Association
NFM	New Faculty Majority
NLRA	National Labor Relations Act
NLRB	National Labor Relations Board
OAH	Organization of American Historians
OSHA	Occupational Safety and Health Administration
PATCO	Professional Air Traffic Controllers Organization, AFL-CIO
PERB	Public Employment Relations Board. The combined Board that now administers all public-sector collective bargaining laws in California. Other states where there is collective bargaining in the public sector have similar Boards.
PhD	Doctor of Philosophy, the terminal degree in most liberal arts and sciences
PSC	Professional Staff Congress
RAFO	Roosevelt Adjunct Faculty Organization
RIF	Reduction in force
ROTC	Reserve Officer Training Corps
SDS	Students for a Democratic Society
SEIU	Service Employees International Union, sometimes affiliated with AFL-CIO
SNCC	Student Non-violent Coordinating Committee
SSDI	Social Security Disability Insurance
SSIs	Service salary increases
STUNAM	*Sindicato de Trabajadores de la UNAM*

SWOT	Strengths, weaknesses, opportunities and threats
TDU	Teamsters for a Democratic Union, reform movement within IBT
TIAA-CREF	Teachers Insurance Annuity Association and College Retirement Education Fund
TSAs	Tax-sheltered annuities
TWLF	Third World Liberation Front, active in the SF State and other student strikes
UAW	United Auto Workers (The International Union, United Automobile, Aerospace, and Agricultural Implement Workers of America)
ULPs	Unfair labor practices
UMWA	United Mine Workers of America, sometimes affiliated with AFL-CIO
UNAM	*Universidad Nacional Autónoma de México* (National Autonomous University of Mexico)
UNITE	Union of Needletrades, Industrial and Textile Employees, sometimes affiliated with AFL-CIO
UPC	United Professors of California, the American Federation of Teachers affiliate in the CSU system until it was disbanded in the 1980s
USLAW	United States Labor Against the War
USW	United Steel Workers
WTO	World Trade Organization

Introduction

These days it's easy to imagine global disaster. It's harder to imagine how to build a sustainable, safe, equitable society. However, that's what we have to do. More specifically, we have to visualize the role to be played in that transition by higher education faculty. The changes we see taking place today, in the wider world as well as in higher education, are accelerating trends that have been in motion since the late 1970s. Foremost among them is increasing inequality, both economic and racial, which is taking place in the context of the climate crisis. These trends have serious political ramifications.

This book focuses on workers in higher education. Although we are college and university faculty, we are also precarious or contingent workers—in the sense that our jobs are *contingent* on factors that have nothing to do with the quality or importance of our work. Our employment lacks the necessary rights and conditions to make the best education for students or to provide a decent life for ourselves. We are now the majority, between two-thirds and three-fourths of all faculty. Our industry is in a crisis. Since the pandemic began, thousands of us have been laid off. Universities and colleges have sent students home (sometimes to other countries), moved to online classes, dropped traditional grading and assessment systems, paused admissions processes and passed emergency rules to allow layoffs of all faculty, including tenured faculty.[1] Even before the pandemic, public institutions faced huge deficits that challenged our system of funding but now many small institutions have checked their balance sheets and decided to fold completely. At the same time, new flexible employment strategies that build on Internet communications are emerging. All of them assume contingency, gig work, as a fundamental feature.[2] This is to say nothing of defaults on the debts incurred by our students, which can only be paid if those who owe have jobs, and will likely trigger a chain reaction similar to the defaults on real estate in 2008.

Now we are faced with a sharp choice. Either we submit to what Naomi Klein calls the "shock doctrine" and open the door to disaster capitalism, or we wield our power to make this social and political crisis deliver something positive for us. In order to do this, we need to think strategically.

1

STARTING WITH A HISTORY

To construct a picture of what can be done, we begin with the history of faculty organizing in the California State University (CSU) system starting (though not from scratch) in the early 1970s. We follow it through into the present, showing the many alliances that were built and the different fights on different terrains that eventually produced what is considered by many to be the best contract and the best working conditions for contingent faculty in the United States. Most of this history took place before there was a real contingent faculty movement, before we used the term "precarious" to describe employment, and before there were lively Internet connections. During this time, faculties were experiencing casualization without really understanding its significance. We make the point that this long fight played a major role in the development of what is now a national and in fact international movement of contingent faculty. We know there will be a new normal: This fight is over what that new normal will be and who will decide it.

Despite the immediate crises, we think that the key characteristic of this struggle is its continuity. A summary update on the 2021 situation in the CSUs is that the struggle has continued, both vis-à-vis the employer and within the union. The Lecturers' Council of the California Faculty Association (CFA) continues to be the locus of debates, the Lecturer leadership has become more diverse, contingent faculty have taken leading roles throughout the union, the union itself has maintained a high level of internal organization and membership despite the loss of agency fee funding, and the contract continues to be likely the best in the US. More good news is that other contracts at other institutions are rising toward this standard.

Today we are still fighting for the same fundamental conditions we fought for in the 1970s. Our working conditions are still our students' learning conditions. It still takes a long, long time to build solidarity. Our arguments about organizing, who should be organized, and how, still hold, with their implications for how we relate to our students and the rest of society. Our definition of tenure as "just cause dismissal" has not changed: just as in other workplaces, employees in higher ed should presume continued employment unless there is a "just cause"—a good, transparent, non-discriminatory reason (and there are legal definitions of this)—for firing someone. Our arguments about the general social wage and bargaining for the public good have not changed. Nor has our "Blue Sky" vision of what a union can mean for faculty.

HIGHER EDUCATION:
A PROFIT-SEEKING INDUSTRY THAT HAS BECOME A BUBBLE

In the last forty years, but especially in the last twenty years in the US, we have seen higher education transformed into a profit-seeking industry. This goes beyond the for-profit sector itself (examples are the University of Phoenix, Kaplan, Grand Canyon University, and the various Academies of Art, to say nothing of the fraudulent Trump University). It has also reshaped the public and non-profit sectors through the ratcheting-up of the cost of tuition, the privatization of student loans, the recruiting of overseas students to stuff sagging budgets, the competition for rankings and the promotional displays of luxury accommodations and recreational facilities. The flow of money through the whole project of academic research has distorted what is studied, what is judged, what is published and who has access to it. The higher ed industry, like the real estate industry and its sibling, the finance industry, has found a way to suck down the wealth accumulated by the previous generation during the 1950s and 1960s, the years post-World War II when inequality for a while actually leveled off and conditions for the working class improved. Many people have noticed that there is a higher ed bubble, a bubble that, when it collapses, will hurt most those who have bought into it, trusting that another credential will make a difference in their employment expectations. This pain will be felt the most, as usual, by those who can least afford it: both teachers and students.

So what remains of higher education when selling diplomas is no longer a quick way to turn a profit, when no number of credentials can get a graduate a job in an economy where there is between 15 and 30 percent actual unemployment, where universities and colleges are stripped of tax support and parents are challenging the price charged for online classes? What does "higher education for the public good" look like in this day and age?

OUR PLACE IN THE LABOR MOVEMENT

We feel it is important, especially at this time, that the contingent faculty movement, of which this book is a product as well as an intervention, sees itself as part of the broader labor movement in education and part of the larger working class as a whole in the US, and in fact internationally, since our core issue of contingency is a major worldwide issue and concern. We hope by this extended case study of one very deep, broad and long-lasting experience in the CSU system to demonstrate that our contingency is in many ways proto-typical of what the future may have in store for other workers. We also think that the story we tell and the lessons we draw may have some usefulness well beyond the borders of precarious academic employment. As highly educated professional workers who are only now coming to realize individually and collectively our place in the working class, we can be an example for the rest of the labor movement. We also of course have a lot to learn individually and collectively as a movement from the rest of the working-class movement in all its variety.

HOW THIS BOOK IS ORGANIZED

We have divided this book into five parts. In Part I, we begin with Chapters 1 through 4 telling the story of organizing within the CSU system (23 campuses, 28,000 faculty, 70 percent of whom are employed as contingent in 2020) that started in the 1960s and 1970s, alongside the Civil Rights and anti-war movements and erupted into what was viewed as a revolution within the union in the late 1990s, finally reaching a point where the California Faculty Association (the CFA, with various affiliates) could bargain what is recognized to be the best contract for contingent faculty in the US.

Part II consists of only one chapter, Chapter Five, that gives a history that goes back to when steel manufacturer Andrew Carnegie brought about the standardization of existing higher education institutions, so that a BA degree from one institution was more or less the same as one from another institution, in what we call the first transition. Readers will recognize the origin of "Carnegie units" at this point. The second transition came about when higher education was impacted by World War II and the subsequent GI Bill. This expanded higher education enormously and also brought a new working-class student demographic through the doors of colleges and universities. The third transition (the 1960s and 1970s) took place when this new student body and their children, much more diverse and working class than in previous generations, changed the curriculum, the demographics of the faculty and the shape of the institution itself, demanding ethnic studies, affirmative action and, central to our concerns, unionization. The fourth and most recent transition began in parallel and partly in reaction to the third, as the neo-liberal agenda beginning in the 1970s worked its way into colleges and universities, public as well as private, leading to declines in public funding, rises in tuition, privatization of institutional functions, vocationalization of programs, and above all, the casualization of faculty, which continually increased, until contingent workers are now the majority of all faculty. We propose that casualization was a solution to a four-part problem confronted by lower-level higher education managers: budget cuts, the uncertainty of enrollments as the student demographic changed, the threat of unionization, and the entry into the faculty labor pool of more and more women and people of color.

This fifth transition is now our challenge. If we agree that the crisis of higher education that we are seeing today is largely the acceleration of past trends, we can expect higher education to rise again because it is needed—essential, in fact. But what will it look like? The fifth transition we want is something we have to build out of that crisis, focusing on how we answer the fundamental questions about our work: for whom, by whom, and for what purpose?

The three chapters (6, 7 and 8) of Part III focus on the union contract and our working conditions generally. Chapters 6 and 7 give a "Blue Sky" vision of what we fight for, that applies to all contingents. We ask, in these two chapters, not what we can get now, but what we need. Concretely, this means looking past the fight immediately at hand, the fight that is constrained by the actual conditions under which it takes place, and visualizing the struggles that will come up in the future. At the same time, we ask what we need to do to make our own organizations more equal and therefore stronger.

In Chapter 8, we turn to the actual CFA/CSU contract to see how these Blue Sky goals map out when they get passed through the give-and-take sausage-making of negotiations with an adversary. We go from Blue Sky to "what we can get." This contract is to our knowledge the best contract for contingents in the US, but we see how language can skirt but still cover an issue (seniority) and how a core topic (academic freedom) can become almost, but not quite, invisible. We will also see how hard it is to bargain a bridge from contingent status to tenure itself.

In Part IV, Chapters 9 and 10 are about strategy. We pick up from the history at the end of Chapter 4 and describe how local and state-level contingent faculty activism took shape as a national movement in the late 1990s. We identify two strategies—the Metro strategy and the Inside/Outside (I/Os) strategy—both arrived at by a combination of process of elimination and common sense, plus some extended discussions among contingent faculty leaders as we attempted to theorize what was going on. We use the story of the Mississippi Freedom Democratic Party to further illustrate what the I/Os looks like, and this brings us into the present. Now with a national movement to push us forward, the urgency of the transition going on around us, and the conflicts continuing to sharpen, we have to look ahead.

Part V consists of seven chapters, Chapters 11–17, addressing what we call "troublesome questions." These questions are intended to be applied in the present, as activists and local unions organize themselves, and are reflective strategic questions with no easy answers. They surfaced during the long, difficult CFA organizing effort, and can be counted on to come up in all unions and union organizing campaigns. The failure to deal with them openly and honestly can undermine the solidarity necessary to win. Because we have been part of innumerable discussions of these issues, we draw on our experience to suggest how to navigate them and what further questions to ask.

The single chapter, Chapter 18, of Part VI opens with a list of the dangers and hopes that we face today. It may seem remarkable that, given current trends, the list of our hopes is so long. We just have to remember that our situation is man (and woman)-made: we made it; we can make it better. The song says, "Without our brain and muscle, not a single wheel would turn." We end the chapter with a few suggestions about immediate strategies.

HOW THIS BOOK CAME TO BE

Readers will see that the first four chapters and many of the examples in the later chapters are taken from interviews with John Hess, who taught in the CSUs starting in the middle 1970s and became a CFA staffer in 2000.

John was a good friend of Joe Berry. Born in 1939 in rural Pennsylvania to a Mennonite family, he studied business and engineering at Lehigh University, was drafted and, despite his Mennonite background, went into the peacetime military and wound up in Germany. He got away from the base a lot, learned German and married. Between his interactions in working-class bars and taverns in Germany and the influence of the Civil Rights movement back in the US, John came to see himself as a socialist and anti-imperialist.

He came back to the United States with his wife Judy and son Andy, enrolled in a graduate program at Indiana University in literature and made friends with people drawn to the new-to-America discipline of critical film studies. This group, which included Julia Reichert and Chuck Kleinhaus, became the founding collective of *Jump Cut*, one of the first serious American film studies journals. His marriage ended in divorce, his wife took their son back to Germany, and John, deciding that it was more fun to be a film journalist than a graduate student, left without finishing his dissertation and went to the Bay Area where he began teaching, first at Sonoma State, then at San Francisco State, and building a local collective around *Jump Cut*.

Joe Berry was a student of John Hess at S.F. State. They became friends. Later they were both delegates to the San Francisco Labor Council. When they both retired, they would meet in John's back garden in Oakland and reflect on their experiences in the world of academic labor organizing. Their conversations led them to decide to write this book.

Joe began a series of interviews with John, taping the interviews. Then John was diagnosed with Parkinson's. After a while, he lost the ability to gather his memories coherently and eventually even to speak clearly. The meetings continued, however, until his death in 2015. The book project was by then well under way, so Helena Worthen, Joe's partner and herself a contingent faculty activist, writer and organizer, stepped in and helped transcribe the interviews and ultimately draft the book. (At this point, we'll shift to the pronoun "we," meaning Helena and Joe.) We contacted others who had lived through the long fight and whose memories could fill in on events that happened back in the 1970s. They gave interviews that shaped the book as it came to life.

Like John, we are socialists and view higher education as part of the social wage of any decent country. Education workers provide an essential part of social reproduction and, like most social reproduction workers, are unpaid or underpaid, and disproportionally female. Of course, like labor itself, education should not, in a just world, be a commodity—an article of commerce—and neither should be healthcare, the environment, the survival of other species, or science. But we should not underestimate the power of the forces that see things differently. Disregard for human life and for the survival of the planet itself has been displayed by the US government. This power will not only resist any challenge, it will move against us in advance, to claim any social space that has opened during this particular crisis and to prevent us from moving forward ourselves. Therefore we have to make the first moves ourselves—or, in labor relations lingo, we must be the moving party.

PART I

THE CASE OF THE LECTURERS
IN THE CSU SYSTEM

1
Student Strikes and Union Battles

John Hess dated his involvement in the battle to a phone call from someone he didn't know:

> From my point of view, it started in 1978 or 1979. I could figure out the date if I had to. I had just begun teaching as a part-timer at San Francisco State. And this guy called me up on the phone, introduced himself as … Bill … He said, "They're going to disenfranchise the Lecturers. It's at four o'clock today in the faculty club. They're going to disenfranchise the Lecturers!" And I didn't even know what a Lecturer was. He sort of went on and on about this. He didn't explain very much. He just sort of emoted. How he got my name and why he called me I don't know.

John was in fact a Lecturer himself, the term used in the California State University system (CSU) for someone who was hired off the tenure line on a per-class, per-semester basis—that is, a contingent. He was one, but he didn't know it. He is telling the story of his first contact with his own union, what was then the San Francisco State chapter of the statewide United Professors of California (UPC), affiliated with the California Federation of Teachers and American Federation of Teachers.

John didn't grasp what "disenfranchise" meant, either. It had to do with what might happen as a result of a law that might soon be passed. People were asking who would be covered by that law. Would Lecturers be covered? Plenty of people did not want Lecturers covered. "Disenfranchised" in this case could mean "not covered" by the law.

This law, if it passed, would be a collective bargaining law that covered public-sector university employees in California. Without such a law, management did not have to bargain with a union. Thus, as far as labor relations in higher education in California went, it was an "eat what you kill" kind of Wild West environment: you got whatever agreements from management that you could force based on your strength on the ground. However, even without a law that required management to bargain, people had formed a faculty union. (Some people might say that a union was even more necessary under these conditions.) It was a meeting of this union that John was called to attend.

At that time there was, of course, the federal collective bargaining law known as the National Labor Relations Act of 1935 (NLRA), which granted and protected the collective bargaining rights of private-sector workers in all states, but state universities are part of the public sector. So the NLRA did not cover public employees (which meant government at all levels) anywhere. The NLRA also excluded many other workers, such as domestic and agricultural workers. Passing laws to enable collective bargaining in the various levels of the public sector was left up to the states. California was just beginning to pass these laws in the late 1970s.

John mainly understood that there was some kind of emergency that he was being summoned to get in on:

This was my first year of teaching at San Francisco State—my second semester or maybe the first semester of my second year. Bill went on and on, and finally, in order to get him off the phone, I agreed to come. Up until that moment I had wisely surmised that it was a good idea to stay out of other people's business. Come, teach your class, and leave! Because if you don't do that, it would inevitably piss somebody off. And given the power imbalance, like job insecurity, it was not a good idea to piss people off. Eventually Bill—Bill Compton, that was his name—he taught math— did piss people off and basically got fired.

But anyway, I went to this meeting and it was packed.

There are several reasons for starting with these quotes. People who work as contingents in higher education will recognize the first: a sense of risk. John figures out that he is being asked to come to a meeting where other people may vote on a proposal to eliminate some of his potential rights. But he knows instinctively that showing up at that meeting to vote against that proposal could mark him as a problem and may cost him his job. Second, it shows that John, who within a few years would become a lead organizer for

Lecturers, started out not really even knowing what a Lecturer was. This is common; people can become great organizers even if they start out thinking that they know nothing. And third, he gets involved because someone asks him to. The personal call makes the difference.

"I don't know how he got my name," John says. He only knew this person by reputation as a troublemaker, but the call got his attention and he went. This is the beginning of John becoming an activist and organizer on his own behalf. Previous experience had already led him to activism on behalf of others, for social justice generally, and even socialism. He had interviewed film-makers in Cuba and East Germany and had gone to El Salvador and Nicaragua on solidarity trips. But this was the start of his becoming an organizer for himself in the place where he could be most effective (and would be most at risk): his own job.

BACKGROUND: THE SAN FRANCISCO STATE STRIKE, TEN YEARS EARLIER (1968–69)

The fact that this meeting took place in the 1970s at San Francisco State University is important, but appreciating its significance requires some historical background.

Under the 1960 California Master Plan for Higher Education, tuition in the CSU was free, as it was in the community colleges and the University of California system. (There are three higher ed systems in California: the community colleges, the CSUs, and the University of California system.) Students neither paid money nor went into debt to take classes in any of these, though they still had to pay for living expenses and books. The student demographic was therefore beginning to include many people whose parents and grandparents had never attended college, especially women and minorities.

Higher education up until the late 1960s had always been predominantly a middle- and upper-class white male domain. Everything from curriculum to student services reflected that fact. A major transition would be necessary to make it a place where this new demographic could get their education. This transition would not be easy. The San Francisco State strike (1968–69), the biggest and longest of many such student strikes during that period, against racism, the Vietnam War, and for students' rights and other issues of decision making at the level of institutional governance, is an example of how hard it was.[1]

13

What some historians now call "disturbances on campuses" were taking place all over the US, Europe, Latin America, Asia and Africa during that decade, making the year 1968 synonymous around the world with student and popular uprisings. The strike at SF State was led by the Third World Liberation Front (TWLF), a coalition of ethnic student groups: Black, Chinese, Latino, Pacific Islander and Filipino, supported by the mainly white chapter of Students for a Democratic Society (SDS) and later, by AFT (American Federation of Teachers) Local 1352, which included many of the college's faculty members, including Lecturers. As we noted above, at this time there was no formal bargaining and therefore no formal bargaining unit. But like the TWLF and SDS, the AFT local was led by conscious leftists who sought radical university reform and internal university democracy with a "serve-the-people" ethic toward the surrounding communities.

So at SF State in the late 1970s, when John got this phone call, the 1968–69 strike was still a recent memory and a point of pride. Novelist and poet Kay Boyle described her experience teaching at San Francisco State during the strike in her 1970 book, *The Long Walk at San Francisco State*.[2] These disturbances, involving the arrests of hundreds of students and faculty, had been going on for over two months by the time she wrote the following in her diary:

> Today, students and police clashed in the most violent campus battle since Columbia [University in New York where students occupied buildings]. The students had called a rally at the speakers' platform, and the police poured onto the campus, broke ranks, and like madmen rushed the students again and again, clubbing them to the ground. (Boyle 1970, p. 50)

The president of San Francisco State, Robert Smith, resigned. S.I. Hayakawa, famous for wearing a jaunty powder-blue beret, replaced him. Photos abound that show the famous moment when Hayakawa climbed onto the strikers' sound truck and tore out the wires. The impact of both that action and the picture redounded to his benefit with Governor Reagan and the right wing, such that he was later elected to the US Senate as a Republican. It also became a symbol of the extent to which administrators would go to repress the movement.

Hayakawa stood on the roof of the Administration Building and shouted through a public address system: "If you want trouble, stay right there, and you'll get it!" Boyle continued:

The students were enraged by the unprovoked attack and more than 2,000 refused to go. They fought the police with everything they could lay their hands on; chairs from the cafeteria, table legs, rocks, garbage cans. By evening, the entire leadership of the black community—moderates, liberals, radicals—had broadcast their shock and anger and announced they would be on campus to defend their children from police brutality. (Ibid., p. 51)

The student strike lasted five months, ending in March 1969. AFT Local 1352 and the trustees of the university also agreed to a settlement, which included many of the students' strike demands.

From the perspective of ten years later, when John started teaching at San Francisco State, that settlement was considered a win, and is still considered so today. The agreement led to the creation of the first and only School of Ethnic Studies in the United States, including a Black Studies Department, many hires of minority faculty, and increased admissions and support for minority applicants as students. By 2020, this School of Ethnic Studies had become a college awarding BA and MA degrees, and had just celebrated its fiftieth anniversary with a remembrance of the strike. Its creation, and the support given to the struggle by the faculty and AFT 1352, also gave a great boost to the movement for faculty organizing.

But from close up, writing in 1969/70, Kay Boyle concluded, "We could not win. For we were opposing a force that goes far beyond the limits of one college president, one campus, one state. We were opposing a nation's fear" (ibid., p. 63).

She was right about the scale and intensity of the opposition, but that's not the same as losing. As long as the fight continued, the war had not been lost. In fact, the struggle, and the partial victory that was won, was a major factor in pushing forward the student movement and the faculty movement nationally as the new decade began. All this was still alive in the consciousness of people who taught at San Francisco State in the late 1970s.

"ONE BIG UNION": WHAT DID THAT MEAN?

John said: "I went to this meeting and it was packed, wall-to-wall professors arguing about "One Big Union."

"One Big Union" is a phrase coined by the Wobblies—the IWW or Industrial Workers of the World—who believed that all workers should be organized into One Big Union. This is a concept that will resonate with contingents everywhere. In this case, it was about who would be included in the potential bargaining unit or even in the proposed law at all. Would Lecturers be in it, or would they be "disenfranchised"—left out?

The 1978 meeting at San Francisco State to which John had been invited reflected the rising awareness among faculty of this new bill being promoted by higher education unions in the state legislature, that would allow collective bargaining in the CSU system. The year before, a different law had already been passed that allowed collective bargaining in the two-year community colleges and the public schools. That law mandated that public employers bargain with the democratically selected representatives of workers; it forced employers to come to the table. Faculty knew that a similar law to authorize collective bargaining in four-year colleges and universities was being worked on, so union activity was starting to bubble up.

Lecturers had a high stake in what kind of representation they might get under this law. Under the concept of "exclusive representation," there could only be one union representing "the same kinds of workers."[3] So, were Lecturers and tenure-line professors the "same kinds of workers" with "a community of interest," which is the legal term? This was like asking if tenure-line faculty and contingent faculty are both faculty, a question most of us have debated at one time or another. If they had a "community of interest" with tenure-line professors, they could be together in a single bargaining unit. If a contingent worker was defined as a different kind of worker, and were split off into a separate bargaining unit, then what about librarians, counselors, coaches, and other academic workers? Would they be considered faculty and included in that bargaining unit, or would they also be separated and have to bargain separately?

The argument at the union meeting about "One Big Union" revolved around this issue, which was really about what kind of union should United Professors of California (UPC) be. For the Wobblies, One Big Union meant organizing longshore workers, farmers, lumberjacks and factory workers all together.[4] Should UPC be something like that: organizing all the workers in the higher ed industry together? Or should it be something less than that?

The statute itself, once passed, ultimately affirmed the right of Lecturers to be recognized as faculty and to bargain, but debates about the "same kind of workers" and the "community of interest" are always heated enough to deserve another look. Another way to express the heart of this debate is to compare it with how craft unions are set up. The building trades unions—electricians, plumbers, carpenters—are craft unions, divided up depending on what craft they perform.[5] Each craft bargains for itself. Were Lecturers a different craft within higher ed? Given that everyone in the debate had a claim to faculty status, how was the craft different if you were a Lecturer rather than tenure-line faculty? So the debate was really about how narrow the definition of the craft should be.[6]

Even in the 1970s, it was clear that if current trends continued Lecturers would soon be the majority of faculty. Although Lecturers' pay had been tied to the same salary schedule that covered full-time tenure-line faculty ever since the CSUs came together as a system in the 1960s, there were limitations which meant that Lecturers were cheaper, especially with regards to benefits. This created a strong incentive for administrations to hire more Lecturers. If Lecturers were included in the union, they might become a powerful force. How would they vote? What if they wanted a bigger share of the pie? Would they take over the union? Could a union led by faculty who lacked job security be trusted to bargain effectively or would they make concessions out of fear? All the issues of status, power, numbers, and, of course, the very quality of education that both trouble and inspire faculty organizing today were in play in that meeting where faculty were arguing about One Big Union.

That UPC meeting John attended would have been full of long-time AFT members. Some of the AFT locals that made up the statewide UPC dated back decades, some even back to the days when teachers first organized.[7] AFT locals had formed on different campuses, spoken out in favor of collective bargaining, and tried to "act like a union," even without legal or management official recognition.[8] The word "united" appeared in the UPC name because, when it was formed in 1970, it pulled together those existing AFT locals plus an existing independent organization of state college faculty, the Association of California State College Professors—thus, "united professors." The UPC was strong, though its members were not a majority of the total faculty. It had a rich labor history. These AFT members were generally in favor of One Big Union.

Other groups within the faculty, however, would have liked to see Lecturers excluded from the new law's collective bargaining rights, or failing that, placed in a separate bargaining unit from tenure-line faculty.

John was right at home in this meeting: "So there were these impassioned speeches. I was a socialist, and some of this sounded good to me."

THE NEW LAW ESTABLISHES A STATEWIDE UNIT
AND SETS UP AN ELECTION

The law that stirred up the faculty at the meeting John attended was passed in 1979. It was called the Higher Education Employer-Employee Relations Act (HEERA). It gave faculty the right to have a collective bargaining agent—a union. It said that Lecturers were faculty and could be represented by a union. It would be a statewide unit, which is what the HEERA legislation defined as appropriate.[9] HEERA left open the possibility that there might be bargaining units for different kinds of workers such as tenure-line faculty, Lecturers, librarians and counselors. A newly established Labor Board, set up to implement HEERA, would settle that later.

Now the faculty had to choose a union to be their bargaining agent—if any—because one option was "no union". The choice would be made through an election. Two unions competed against each other in this election. A major issue that distinguished between them was which one was more likely to want to include Lecturers in a single bargaining unit with tenure-line faculty.

One choice was the UPC, the one John was involved with. It was affiliated with the California Federation of Teachers, the American Federation of Teachers and the AFL-CIO. These affiliations marked it as part of the labor movement. The UPC had historically been referred to, and perceived as, the faculty *union*, as opposed to one of the various *associations*. It was the choice more likely to want to include Lecturers with tenure-line faculty.

The other choice was the Congress of Faculty Associations (CFA). (Later, the same acronym would be used for the successor organization, today's California Faculty Association.) The CFA was a grouping of four statewide *associations* that had presences on some campuses: the California Teachers Association, affiliated with the National Education Association (CTA/NEA); the American Association of University Professors (AAUP), which was also in the CSU system; various other unaffiliated associations and later, the California State Employees Association (CSEA), the independent association of nonteaching employees. The term "association" was intended to distinguish these groups from a "union." These were conservative organizations that had historically opposed collective bargaining in principle in favor of "professional consultations." They did not see themselves as part of the organized labor *union* movement. If they were going to be forced into collective bargaining at all, they would prefer to represent only tenure-line faculty.

So the CFA and the UPC both campaigned to be the bargaining agent for the faculty. The essence of the campaign run by the CFA—the *association*—was "Stop AFT!" By this the CFA meant: *oppose* having a collective bargaining agent that was affiliated with the non-education labor movement, and *block* any possibility of an organization that could ever be largely controlled by Lecturers, whose numbers were rapidly increasing. This is not very different from the warnings being passed around today against allowing the US Democratic Party being taken over by the "radical left."

The UPC campaign strategy, unfortunately, was mainly defensive. They tried to ignore the whole issue of Lecturers and forestall the tenure-line faculty's fears of being taken over by them if UPC won.

The Lecturers could see that this was a losing strategy. Lecturer leaders sent specific proposals warning that the UPC would lose unless organizing among Lecturers was prioritized, but they were ignored. It seemed almost hopeless. John asked around to see what he could do in addition to coming to meetings. He was told that no one was actually representing Lecturers at San Francisco State at the Delegates Assembly (DA), the governing body of the UPC. When chapter presidents went to the DA, they had to bring someone. John said, "On many campuses that slot was empty. So they would just say, 'Hey you!' A lot of other people who did appear were 'Hey you!'"

So John and two others—Mina Caulfield and Bill Compton, the man who had called to bring him to that first meeting—started going to the DA and Lecturers' Council meetings. Lecturers and other contingent faculty who are involved in organizing today will recognize John's description of what went on: "They were down in these shabby hotels in Los Angeles and San Francisco. These meetings all had the same sort of format. We got in there and sort of mixed it up with people that I would come to know well … Progress? I don't think we made a lot of progress. But we made a lot of noise."

A lot of the noise was expressing frustration. A brief history of this period written soon afterwards by Lecturers relates that they believed that their conditions made it difficult to have "even the barest minimum of professional integrity."[10] But suffering was not the same as organizing.

Lecturer leaders started a desperate fight to persuade as many people as possible to sign up as members of UPC and gain their support in the upcoming collective bargaining election. Elizabeth Hoffman, who twenty years later would be on the union negotiating committee, was a Lecturer at Cal State Long Beach and remembers climbing up and down stairs in different buildings, conspicuously pregnant with her daughter, trying to find Lecturers and sign them up for the UPC. "I thought, people must feel pretty sorry for me," she says.

Bad feelings ran deep. The CFA made no secret of looking down on Lecturers and hoping to split them off into a separate bargaining unit. Susan Meisenhelder was a Lecturer and had taken on the Lecturer representative position for San Bernardino State. She taught English with a specialty in African American and Women's Literature. She said: "When the CFA people came on campus to talk to people about why they should vote for the CFA, their rap was, 'Lecturers aren't our kind of people.'" She was sensitive to the innuendos of the phrase "our kind of people." Twenty years later (1999) she would become president of the union.

By this time Lecturers as a group increasingly included African-Americans, Asians, Latinx and women, including some radical veterans of the movements of the 1960s. These hires were the result of affirmative action goals pressed by the strike in 1968–69. This contrasted with the make-up of the tenured faculty who, as a whole, were more white, male and politically conservative. Realizing this, in 1975, a small group of people trying to organize Lecturers into the UPC had sought out the Statewide Affirmative Action Committee to get some advice. Members of this mostly minority faculty union committee, which had itself been formed to fight inequality, drew on their experience to advise the Lecturers to organize themselves first and then seek official representation as a distinct body within the union. (This is an example of the Inside-Outside strategy we will describe later.) The group followed their advice and, eventually, this would lead to getting a seat on the UPC State Executive Board for a Lecturer and an official Lecturers' committee, which eventually became the Lecturers' Council. But at the time, the union leadership did little to help, and what they did do was not enough to ultimately win the election for the UPC.

TWO ELECTIONS: UPC WINS THE FIRST ONE
BUT THEN LOSES THE RUNOFF

In the first representation election, in 1982, faculty were presented with three choices: the UPC (the *union*), the CFA (the *association*), and "no rep." Voting "no rep" would mean that the voter did not want representation by any bargaining agent. As expected, the highest number of votes came in for the UPC, which got a plurality (over 40 percent) but not a majority (over 50 percent) of the total vote. It was good, but not good enough. Next came the CFA, and last, with relatively few votes, came "no rep." The failure of one organization to get a majority meant that there had to be a run-off between the two top choices. "No rep" would not be on the ballot in the run-off, only the UPC and the CFA.

Not surprisingly, the no-rep votes, the people who didn't want any union at all, voted for the more conservative CFA. In the run-off, the UPC was defeated, but narrowly, by less than 50 votes out of 15,000 possible votes. Jack Kurzweil, who taught electrical engineering and was openly a Communist, was the UPC Chapter President at San Jose State at the time. He hosted the UPC campaign office in the garage in his backyard ten blocks from the San Jose State campus. In an interview in 2019, Jack Kurzweil explained why the UPC lost the vote:

> There were three reasons why UPC lost. One was [AFT President] Al Shanker's racism, which drove votes away.[11] In fact, the School of Ethnic Studies at SF State went for CFA out of opposition to Shanker. Second was the narrowness that AFT exhibited, through its staff, of its vision of the campaign—a vision that was narrow. Third ... the failure to prioritize campaigning among the Lecturers for fear of losing what was in fact a very small number of tenure-track votes. So Lecturers voted at a much lower rate than expected. This more than explains the tiny difference in the final vote.

In other words, the fear of alienating tenure-line faculty by conducting a strong campaign among Lecturers may have cost the UPC that victory. This distinct defeat for the Lecturers was a shock, and left many people frustrated and angry.

There was a window of time when trying to decertify the CFA would have been legally possible.[12] Many faculty angrily urged a decertification campaign. Jack Kurzweil made a private backchannel attempt to try a merger instead:

> After that election, which CFA only won by 47 votes, UPC was hell-bent to hold a decertification election.[13] I had a discussion with Bill Crist, who was President of CFA, in which Bill and I talked about a merger ... Crist was amenable ... and I presented it to UPC. I was roundly denounced![14]

The problem, which was a real one, was that he was not authorized to have such a discussion. But Kurzweil also remembered:

> Not long after that, John Hess in a very quiet way said to a meeting that he found CFA's approach to including Lecturers in the bargaining process superior to that of UPC's and that he was prepared to work with CFA on that. The thing that I found remarkable was the way people responded to him. In contrast to me, they responded to him with just great respect. I, on the other hand, got pilloried! I mention this because it has to do with how the same body responds to people differently. John mentioned his intention to work with CFA on their first contract negotiation; he was responded to with respect. That's an indication of the kind of stature he held.[15]

By then, the new labor board included under the HEERA had been set up, had decided that Lecturers and tenure-line faculty were the "same kind of workers," and put them all in one bargaining unit together. This laid the One Big Union issue to rest at least for the time being. The CFA now had to bargain for both tenure-line and Lecturers as one combined faculty unit. They moved hastily to wrap up a first contract. Nina Fendel, a labor attorney who came on staff in 1988 and worked for the CFA for nearly twenty years, described it as "a kind of inferior contract bargained right away to get one in place in six months."

The speed of bargaining was facilitated in part because, ever since the CSU system had been set up, the employee relations scheme had been based on the State of California civil service regulations. The story went that one of the original trustees of the newly created CSU system was a top manager in the state civil service and never questioned that similar conditions might not apply to academics who were, after all, also state employees. Bargaining could go forward or backwards from what the civil service regulations laid out. In the absence of a contract, the law would rule. As Jack Kurzweil pointed out, later on in the 1990s when there was a threat of a strike, the union could threaten that, in the absence of a contract, they could "revert to" civil service conditions. That was something that the Chancellor's Office did not want, because what was in the law might be better than what the CFA had negotiated.

The decertification issue disappeared in 1983, because, before that first contract had run out, the CFA became affiliated with the AFL-CIO. This happened because their member organization, the California State Employees Association, had newly affiliated with the AFL-CIO through Service Employees International Union (SEIU). That meant that the CFA acquired protection against any decertification because both UPC, as an AFT/AFL-CIO affiliate, and the CFA were now AFL-CIO unions and came under the AFL-CIO constitutional prohibition against member unions raiding each other. This is doubly ironic, because the CFA had heavily campaigned against the UPC on the basis of its labor movement AFL-CIO affiliation. After the election, the CFA, the Congress of Faculty Associations, renamed itself the California Faculty Association or CFA, but retained its three affiliations with the NEA, AAUP and CSEA.

To many readers, the last few pages may seem like a typical who-did-what-to-whom of organized labor, an exercise filled with conflicts and confrontations that make no sense and matter to no one outside the in-group. Those readers, who are probably more interested in getting on with the story of organizing, should try to remember only this: history does matter. These events left scars. For readers who already have some experience in labor struggles, this story may feel as dramatic as a Greek tragedy with its broad, sudden shifts in the weight and position of power.

So now, the CSU faculty had chosen a union as the collective bargaining agent that had been forced to represent Lecturers (which the leadership had opposed) and was now affiliated with the AFL-CIO (a connection it had also opposed), and these contradictions were on full display. Looking back, John Hess sometimes said, "In the long run, the CFA win might have been a positive, because it forced the UPC people in the CFA to organize and ultimately carry out the Revolution in the organization."

Between 1983 and 1999, when that "revolution" finally took place, there was the long decade and a half of economic and political shocks of the Reagan era. Nevertheless, at the price of slow, grinding effort, some substantial gains were made toward what would become the famous "good contract" of the 2000s.

2
Layoffs and Hard Years
for Organizing

In the mid-1980s, Elizabeth Hoffman, whom we last saw climbing up and down staircases in buildings at Cal State Long Beach, hunting for Lecturers to sign them up for the UPC, was a single parent taking care of her daughter. In addition to working as a Lecturer at Cal State Long Beach teaching literature and linguistics, she also taught at one or two community colleges nearby. She, like many who were strong UPC supporters, had vowed never to join the CFA. But one day a CFA staff person found her in her office one day and asked her to join. This would have been soon after the first contract was bargained. "No, no, no," she said:

> Then he said, but it's the only game in town, so he talked me into it and I filled out a couple of membership cards. Then he said would I be interested in running for the Lecturer's position. I had just joined! I said no, there's no possibility, I'm too busy already. So he said, "We just need a name, we already have someone who wants to run, but we want it to be competitive." So well, all right, I could be a name. Then the ballot came out and I was the only name on it. I called up and said, "What's this? I don't want to do this! You said I would just be a name!" and he said, "Well, we're hoping for a big write-in." So I wrote in someone's name but I won anyway. It's hard to lose when you're the only one.[1]

Like other strong UPC supporters, once Elizabeth joined the CFA she became very active. She would be a member of the bargaining team, and eventually become vice president for Lecturers and president of the Lecturers' Council. She would work closely with Susan Meisenhelder and John

Hess. However, her first experiences as a leader were handed to her without much warning.

One day another active Lecturer, Sherna Gluck, called her up and said, "You need to call a meeting":

> So I called a meeting, people came, and when it started Sherna just sat there. I thought, "Why doesn't she start the meeting?" She said, "Well, Elizabeth, aren't you going to start the meeting?" I said, "Me?" She said, "Well, you're the Lecturer rep." She was very much just telling me to do stuff. So I had a meeting. Then she called me and said "You have to have a committee and a meeting at least once a semester." But things picked up. We started getting active.

People began to look to the Lecturers' Committee on campus for help with problems. One of the first long fights had to do with step increases, which are standard increases in salary based on years of service; for example after three years, six years, after ten years, and so on, step by step. This is a widely used way to reward commitment and experience. Elizabeth said:

> At that time Lecturers did not get step increases. The administration even called it "merit pay" to make it look like it was awarded for special merit. Even though the contract said step increases went to all faculty, essentially, you had to be among the tenure-track faculty to get it. Maybe they got them up at San Francisco State, which was always ahead of other places, but at Cal State Long Beach, never. In fact, bringing it up made people mad.
>
> So a Lecturer called me up and said she wanted me to go with her to talk to the Dean because she didn't get her step increase. I had never represented anyone and had no idea what a grievance was. I said, "well, you should call the union and see if a staff person can help you." But people did not know any staff well. So I said, "I'm perfectly happy to go with you," and we went together. Here we are in the Dean's office and she kind of looked at me, and I just sort of sat there. Then she said, "I brought my union rep with me." I kind of looked around. She said, "Elizabeth?" I said, "Oh, yes, you should give her the salary increase." The Dean said, "I can't give you a salary increase. I would be glad to write you a letter about what

26

a good job you're doing." And she says, "The problem is, they won't take that letter at Ralphs." [Ralph's is a grocery chain in Southern California.]

When we came out she said, "You need to start a fight so we can get step increases." I thought, that's the equivalent of saying you need to make sure the sun does not come out tomorrow. But we did start a fight to get step increases, not "merit pay," which is a way to give raises using a whole lot less money, because you hardly give it to anyone. It was a long fight.

So the union's goal was to get the term "merit" replaced by "salary step increase" (SSI) and to get SSIs applied equally and universally to Lecturers. Elizabeth wrapped this long, complex fight, with its multiple stages of member education all into one cheerful sentence:

First people had to apply, some were turned down, then we had to grieve that—it was one of the first things Lecturers did, learned the grievance system—and finally we ended up bargaining an application process—we had to learn how to do that—and ultimately we got automatic step increases and now no one thinks in terms of Lecturers not getting step increases.

These labor-intensive, long-drawn-out fights produced individual wins that brought a few hundred dollars a semester into the pockets of individuals, but slowly they cumulatively contributed to strengthening the union, developing leadership and increasing the stability of Lecturers' lives. Nina Fendel looks back on those early fights:

What caused the Lecturers to turn the corner and begin to organize their own power within the union? It probably had a lot to do with the young leadership, people more sophisticated with building power ... There was nothing but custom and tradition holding them back and they were willing to buck that.

THE *YESHIVA* DECISION AND THE LONG SLIDE

For higher ed unions in general, the Reagan era was a holding pattern with small but numerous retreats. That period saw a major judicial loss—the *Yeshiva* decision—which declared, improperly and against the evidence, that tenure-line faculty in private higher education institutions were management and therefore unprotected under the National Labor Relations Act for collective bargaining.[2] Given the post-PATCO (Professional Air Traffic Controllers Organization) anti-union climate, this decision encouraged many unions in higher education to assume a defensive circle-the-wagons posture.[3] This was in contrast to the aggressive organizing posture they had taken in the 1960s and 1970s that had led to high-density faculty representation nationally, especially in California, New York, Illinois, Washington, Oregon, Massachusetts and Wisconsin, in both community college and university sectors.

Nonetheless, the faculty union in the CSU system managed to make some important gains. One was the step increase struggle that Elizabeth Hoffman talked about, which would actually recur time and time again as the administration attempted to propose differential pay schemes under different names. Another achievement was building a track record of precedent-setting arbitrations to back up the language of "careful consideration." In the early 1970s, before the first contract, there had been a five-years-and-out policy and faculty would get "churned."[4] In the first contract, the CFA bargained that administrators would practice "careful consideration" with reference to laying off Lecturers. This innocuous-seeming phrase turned out to have more power than was obvious at first glance. Again, member education was key. It was possible to actually show whether "careful consideration" had taken place or not. Nina Fendel explains:

Sometimes we would find that the Chair of a Department was laying off people and then bringing in their brother-in-law. Then we would find that the administration had not looked at the official file of the person who was laid off before doing the new hire. That was a requirement! They often kept duplicates of the file in their office but there was an official file, and you had to sign it out in order to look at it. We would look at the official file and see that no one had signed it out. So where was the careful consideration, if they hadn't even looked at the Lecturer's file?

We had these cases not only in times of layoffs but other times as well because there was a lot of monkeying around. There were certain Chairs that thought that people got stale and you should bring in someone new … We would ask, "Did they look at the file?" and then we'd subpoena the papers of the person who had been hired as a replacement and often

we'd find that they had never taught the class. Whereas we'd find that the
person who was laid off had taught the class and had [a] PhD.[5]

Arbitrators who got the case would note that the "careful consideration"
standard had not been met.

A risk of this strategy was that sometimes the Chair and the person
who was filing the grievance—the Lecturer who was laid off—were both
members of the CFA, so the union would appear to be filing a grievance
against another member. The union had set up a Faculty Rights Panel with
a remit somewhat broader than grievances, something like a Steward's
Council in other unions. But Fendel explains:

> We handled the Chair-member problem by having someone from the
> Faculty Rights Panel represent the Lecturer. Say the Lecturer comes to
> the union and says, "I didn't get any classes." So we file a grievance. If
> the Department Chair is a member, we would say, "Chair So-and-so, the
> Lecturer who has been laid off will be represented by X from the Faculty
> Rights Panel. If you have any questions about the process of the grievance
> he or she will answer them for you. We won't actually file the grievance
> against you, it's against the Administration. There will be a grievance
> hearing but you don't have to be there unless the Administration wants
> you there." The administration often did not want them there, because
> they were union members too. So the Chair would not be there at all.

This is a solution to a problem that has faced many combined bargaining
units, where contingent and tenure-line faculty are all members of the same
unit.

CUTBACKS IN THE 1980S, LAYOFFS IN THE 1990S, AS THE ECONOMY CHILLS

Nationally, casualization in higher ed was a trend that had steadily increased
through the 1980s. Along with this increase came private anger and public
criticism. The bad working conditions, lack of job security, low pay, and lack
of opportunities for professional advancement were attracting the attention
of more and more people working in what felt like the lower depths of colleges
and university systems. However, at this point, there was no national, much

less international, network or movement to support a general campaign against these conditions. There were struggles taking place in New York's City University of New York (CUNY) system, and in the California Community Colleges, the rising fight based in the giant CSU system during the 1980s and '90s was the biggest and most important at that time, but these fights were going on in isolation from each other.

The Ronald Reagan era was bad for labor, but for businesses and the stock market things seemed to be good, so it was possible for some people to say "the economy is doing well" and feel things were fine. In 1978, Proposition 13 was passed, placing restrictions on tax increases that would slowly decimate California K–12 funding.[6] In addition, the Cold War ended, proof of American superiority. But then, under the administration of President George H.W. Bush, there was a financial crisis. The 1991 First Gulf War and surges of war patriotism distracted the Bush presidency from responding to demands for the "peace dividend." In 1999, rising internationalist challenges to the New World Order were symbolized by the anti-World Trade Organization (WTO) protests. Then, in 2001, the 9/11 attacks "saved" the presidency of Bush's son George W. Higher education experienced this slow-motion global financial tumble as severe cutbacks, that led to layoffs and the closing of whole programs. In California, there was also a wave of political attacks specifically on teachers to justify these cuts.

Jane Kerlinger, an activist Lecturer who taught Environmental Science at Chico State and would later become statewide executive vice president of the CFA and then CFA staff between 2001 and 2006, remembers that there were so many layoffs at this time that the union just left a stack of grievance forms on a chair outside the union office for people to take home and fill out themselves. These would become "careful consideration" grievances and would eventually be taken to arbitration, but it might take three years for a person to get their job back. The backlog would pile up; eventually there were over 300 incomplete arbitrations waiting.

There were two major waves of layoffs, one in 1991 and one in 1993. Nina Fendel explains why they were able to get people's jobs back in 1991, but not 1993: "The first mass layoffs, they found money. It was set aside in a kind of pot within the CSU system where discretionary funds were kept. The second time that money had been used up and they said they couldn't find more."

There had been budget cuts at the state level. Kerlinger remembers her own layoff in 1991:

I got laid off and filed a grievance against it and won. I won on the basis of retaliation for organizing. I had taken pictures of the Psychology Lecturer's office with about ten people trying to work with one computer.[7] Chico is small—you can really influence public opinion! I was working on starting a water quality lab out of the College of Engineering, and my Dean couldn't stand me. He manipulated schedules and chose someone who had never taught Environmental Science [to] teach it, and I was supposed to train that person! I wasn't going to take that lying down. I got a small monetary amount and my job back next semester.[8]

She remembers that she went through mediation rather than arbitration. If she had gone through arbitration, the decision would have set a precedent that would have applied to other Lecturers.[9] A sense of chill was running through workers in public education across the country. Programs were slowly but surely being cut back, many of which affected not just faculty but students unequally, hurting the less-prepared and first-generation college students more than those who were luckier and better prepared.

THE SOLUTION TO INSECURITY: "GET ANOTHER DEGREE"

An individual's solution to this lack of security was to go back to school and get another advanced degree, in the hopes that such a move would strengthen one's resume. Helena Worthen did this in the 1990s, going back to UC Berkeley in the Graduate School of Education. Although by this time UC had started charging tuition, it was not yet expensive. The graduate student research assistants had organized into the UAW (United Auto Workers) and would soon win recognition and bargain a contract that covered tuition, health insurance fees, and a $900 per month stipend, so that, on top of her other contingent teaching and union jobs, it was do-able. Joe Berry did this in the 2000s, going to what was then the Union Institute to get a PhD in Labor Studies.

John Hess had already gone back to school. He had been ABD (completed All But Dissertation) from Indiana while he was teaching at San Francisco State, so in the late 1980s he finished up his PhD. His dissertation was on Cuban and Italian cinema, drawn largely from his work with the journal *Jump Cut*, hoping that a completed degree would give him some job protection:

31

I had mistakenly imagined that there would be some sort of conversion process of Lecturers into tenure-track positions, and I knew that if there was such a thing, having a PhD would be one of the necessaries. So although I was way over the time limit, I wrote them a letter and they gave me a test—a softball test to write. Then I wrote the dissertation and that was that. But by now it was clear that teaching at San Francisco State was not going to go anywhere for me.

The idea that teaching for years off the tenure line would somehow morph into a tenure-line job was a pipe dream, as many others were discovering. The phenomenon of the permanently temporary was hardening.

Up until 1992, John Hess was still a Lecturer. He would not get hired as a CFA staffer until 2000. But as a Lecturer, he became chair of the CFA Lecturers' Council. This is what had developed out of the old UPC Part-Time Temporary Committee of 1975. It was an official body of elected Lecturer representatives from each campus. He was also on the CFA statewide bargaining team in the CSUs. He had been the last chair of the UPC Lecturers' Council in 1981, which made him the only person to have been chair for Lecturers under both the UPC and the CFA.

He was thus a very visible activist leader at a time when hundreds of Lecturers were being laid off. The second wave of cuts hit him, too, even though by then he was the only Lecturer left teaching in the Film Studies Department. He went to the union to see what could be done. Fendel said, "There was no money. He was the senior guy, very effective, well-liked. But there was just no money." Like so many others, John started looking for work elsewhere.

By then, Helena Worthen and Joe Berry were also looking elsewhere. Between them, they had held eight different jobs. Joe had the only full-time job with benefits, a union staff job which he was ready to leave. Helena was nearly done with her PhD, and held various elected and part-time staff positions with the California Federation of Teachers. One night when both were still at their desks at midnight, they said to each other, "This is not doable. Whoever finds the first full-time job, we'll go there, no matter where it is." It turned out that Joe, with an MA in history from San Francisco State, many years of teaching history and Labor Studies, plus various jobs for the California Federation of Teachers, got a position at the University of Iowa in the Labor Center. Joe took that job in 1994 and Helena followed in 1995 to write her dissertation.

When John Hess went looking for a job he found one even further east, at Ithaca College in New York. It was a full-time, tenure-line position. He would stay there for five years. But before John was laid off at San Francisco State, he had married for the second time. His second wife, Gail Sullivan, had been working in San Francisco as a United Parcel Service (UPS) driver. She was an activist in the rising Teamsters for a Democratic Union (TDU) reform movement.[10] She was the first woman to be elected business agent—a term for union staff representative—in the Teamsters in California. When John took the Ithaca College job, she and John commuted back and forth every month between Oakland and New York. But a few years later, TDU and its allies would elect a new IBT (International Brotherhood of Teamsters) national president, Ron Carey, and Carey offered Gail a staff job. This meant that Gail would leave their house on 65th Street in Oakland and move to Washington, DC, where the national headquarters were located.

By then, John had been at Ithaca five years and would have come up for tenure, but when he heard about Gail's job offer, he decided to join her. He said, "That's good enough for me," resigned from his job at Ithaca and moved to Washington.

Then, only a few months after they got to Washington, the government ousted Carey,[11] who had been accused of using illegal funds to subsidize his reelection campaign. The old-guard Teamster leadership, led by Jimmy Hoffa Jr., was relentless in persecuting him and pushing the government to invalidate the election. Ultimately, Carey was personally exonerated. It turned out that although outside funds were illegally fed into his campaign, he had been excluded from knowledge of it.[12] But the immediate impact on John and Gail was that Carey was out, Gail lost her job and John had no job at all.

So they went back to Oakland in 1999, where they still had their house. Gail went to work for reform Teamster locals in the Bay Area, helping them set up their office systems. Later, she got a job with the California Nurses Association (CNA).[13] Altogether, John had been away for six years. In spite of the cutbacks, he was now able to get a class to teach at San Francisco State.

Helena and Joe, in the meantime, circled through labor and university jobs in Pennsylvania and Illinois and didn't return to the Bay Area until 2010, when John and Joe reconnected and started the work that led to this book.

3

Revolution in the Union

Between 1993, when John left California, and when he came back in 1999, a serious opposition to the old guard had emerged. Nina Fendel attributes this in part to a generational divide, with the younger generation being more openly activist. However, this younger generation was mostly Lecturers or former Lecturers who experienced their jobs in the CSUs as something very different from the jobs of tenure-line faculty, so the opposition to the old guard had a material basis.

Nina Fendel remembers Jane Kerlinger as a young activist who could bridge the divide between the tenure-line faculty and Lecturers: "Jane up at Chico State was someone who was very good with dealing with the older, more tradition-bound faculty, because she had great interpersonal skills ... She just knew how to appeal to the sense of fairness of the older faculty."[1]

At that time, at Chico State, all Lecturers would get layoff notices every year, and then less than a week before classes started, they would get re-hired—or not. The insecurity of this could be incredibly stressful. Lecturers had to turn in their keys after each semester, buy their own parking passes, and had no mailboxes. Kerlinger said:

I first got involved in the union up in Chico because we had a campaign based on "respect for Lecturers." In 1990, I joined the executive committee [of the Chico chapter]. In 1991, I filed a grievance against being laid off and won, on the basis of retaliation for union organizing. I went to the Lecturers' Council for the first time in 1992, and what I saw was the same as what John saw. They still had people who were just complaining about how tenure-track people did not treat them with respect.[2]

She ran to be vice president for Lecturers, which made her also the chair of the Lecturers' Council between 1995 and 1999. In 1999–2001, she became executive vice president in the election that brought in Susan Meisenhelder as president.

In 1994, intensive union lobbying brought the passage of Assembly Bill 211 in the California State Legislature and got it signed by Governor Gray Davis. This gave Lecturers in the CSU system access to health benefits if they were teaching 50 percent of course load or more. The definition of full time was twelve units, so with teaching six units a Lecturer would be covered. The union's work then became informing people of this arrangement and getting them to enroll, which encouraged contact among the members. "We had our program together," remembers Jane. Note that this benefit—access to health benefits—was done by statute, not by bargaining. Then in 1995, the union was able to expand this in bargaining so that they got thousands more Lecturers health benefits at 40 percent.

But the broadening involvement of Lecturers through teaching them about this new benefit led to pushback from the "old guard" in the union. If some Lecturers weren't even dues-paying members, why spend time telling them about their benefits? Kerlinger remembers how she dealt with it at Chico State:

> One of the complaints was, "We're spending too much time on Lecturers and they're not even organized." I took that as a call to arms. You know, Lecturers are difficult to find. They are all over the place. But my lab was just down from my office, so I could go up to my office and get on the phone and sign people. To find out what their office hours were, I would go to the department secretaries and ask them to print out the office hours; I'd say, "I have to find this person who is not a member, so I need to know their office hours and their extension." So I would call them and tell them what we were doing. I would say, "I need you to join and I can sign this for you. If you don't want me to, give me a call tomorrow or the next day, but don't hem and haw, it's just yes or no."

Technically an organizer is not supposed to sign someone else's name, but apparently no one objected at the time.[3] Jane explained:

> There was a line above the signature line for you to report who signed the member up, and I would put my name there. I knew most of the people and they knew my name and they trusted me. It took a while for some of

them to soften. They'd say, "The union isn't doing anything for me." I'd say, "We are the union." We had some fun with that. There was hardly anyone who really said no. It was more a matter of finding people who didn't have offices.

She had a range of approaches at her disposal and used everything she could think of:

I had a newsletter going out into everyone's mailbox. I would take all the flyers to the main room, about half a mile away from my office. Each newsletter had a label on it. I had a dinner party at my house—I had the most prominent Lecturers on campus at it. Recruiting proved that we were getting organized. I was following the numbers very closely. On many campuses we surpassed in number higher than the membership of tenure-track faculty.

THE BAD 1995 CONTRACT

Despite the ground-level work being done by Lecturers with regards to the layoffs in 1991–93 and despite the outreach efforts to inform Lecturers of their access to health benefits after 1994, this organizing did not build strength or reach up to the top levels of the union in a way that made a difference in the next contract. Another way to think about it is that it made a difference in a negative way. The old guard—the mostly white, older male tenure-line leaders of the union—Elizabeth Hoffman described them as "patriarchal"—pushed back passively by simply ignoring the Lecturers' demands. Therefore the contract negotiated by the CFA in 1995 was stunningly unsatisfactory. The worst part was that it sacrificed the two-year appointments for Lecturers, the only form of job security that was available to them. People were incredulous. In fact, the contract was criticized so widely that John, back East at Ithaca, heard about it.

Susan Meisenhelder, by then major faculty leader with a progressive and unified vision of the CFA, talked about the 1995 contract:

I think most people would tell you that the 1995 contract was the turning point. It was just horrible. It wasn't even just from my point of view that it was horrible. It was that there had been no fight put up whatsoever. It was billed as a real "savvy" agreement that would work out in the long run, but it had the worst kind of merit pay and decimated the workload provision—even to a fairly inexperienced person it was a horrible, horrible contract. So that started getting people really rumbling.[4]

The opposition starts to form

Inevitably, a group of angry people began to meet informally and privately and discuss what to do.

People started meeting in secret, talking about the fact that it was embarrassing to be a CFA leader because the faculty had no respect for the union. So it was clandestine for a long, long time. It was a tough, not very fun union experience for a number of years. Many of us stayed with it out of sheer meanness; we weren't going to let it go without some kind of fight.

Probably one of the reasons that the opposition grouping was slow to form in the early 1990s, and even after the 1995 contract, was because these progressive, mostly former UPC members, including many militant Lecturers, were reluctant to oppose the then current president, Terry Jones. He was an African American and came out of the older pre-collective bargaining culture leadership.[5] Elizabeth Hoffman described this leadership as seeing itself as progressive and egalitarian but in fact was condescending to the Lecturers.

The idea of forming an opposition caucus was laid out explicitly by Professor of Government Jeff Lustig, chapter chair at Sacramento State, and head of the Center for California Studies. They met at Jack Kurzweil's house in Berkeley. Susan Meisenhelder was there. She told the group that she was willing to run for president when the next election came along. They all pledged to actively support her but did not go public with this at the time. She would be elected CFA president in 1999.

In the meantime, another bargaining round came along in 1998–99. Bargaining for this contract was very unpleasant. Negotiations dragged on beyond the life of the contract so that people were bargaining while working without a contract.[6] Things were going nowhere. Despite the bad experience of the 1995 contract, faculty had not been mobilized to support the bargaining team. The Lecturers' Council seemed to be out of the loop entirely. Morale was scraping bottom.

But by now the underground opposition caucus was becoming more publicly visible as an active group. Meisenhelder, who was on the bargaining team, foresaw that the union was going to cave and agree to yet another contract, this one just as bad as the 1995 contract. A "tentative agreement," or TA, was being drawn up. Her opposition caucus decided to take a public stand. They were going to publicly declare that the TA was not good enough. As Meisenhelder recalls:

> We could see the same thing coming down the pike. We had done our best to do the fight internally and convince the leadership that we could do better, but eventually we decided to organize a full-blown Don't-Support-the-Tentative-Agreement campaign. We had enough power on the Board of Directors to do what I now realize is the unthinkable, to get them to take a "No Position" position on the tentative agreement negotiated by the bargaining team.

This would mean that the statewide Board of Directors would be asked to send a "No Position" position—basically saying "We do not want to make this decision"—to the Delegates' Assembly (DA) and hand it over to the DA to decide what to do.

The Board itself was split, so the opposition group, taking a big risk but, at the same time, trusting the depth of their support, called for public debates on whether to support the tentative agreement or not. These debates were held on each campus, separately, in person. They were well-publicized and well-attended, even crowded. Microphones were set up so that people could come forward and line up to make their arguments. Meisenhelder again: "Individual leaders were free to speak out independently on what they thought, so people lined up. We had the pros and the cons. That brought out some old—not in years but in experience—leftists, people who had been around in left-wing politics, and that helped a lot."

When it came to a vote in the DA on the tentative agreement, the members of the DA rejected the tentative agreement. She said: "And to make a long story short, it was defeated. This was the first time it ever happened. I remember thinking, 'The easy part is done now. Now the hard part comes.'"

Just as much as losing, winning has consequences. The next part would be hard because the rejection of the tentative agreement had come just as an election for the leadership of the union was scheduled.[7] The old leadership's work had been rejected, and now the next thing would be the creation of a new leadership. But if the union leadership changed, it would mean that a new bargaining team would face the CSU management across the table to finish the 1999 contract. Could they mobilize the membership fast enough to come in from a position of strength, or had they just weakened the whole union's position by discrediting the current leaders?

A CRISIS FOR THE UNION STAFF

Susan Meisenhelder, as the acknowledged leader of the opposition group, decided to go public about running against Terry Jones, the incumbent president. This was going to be very divisive among the union staff. Staff works for the union as a whole and is not supposed to take sides. Yet the old guard of the elected leadership had revealed their inability to convince the members that they were doing a good job. Nina Fendel, as one of the union staff, was worried about what would happen if the election brought in a new, raw union bargaining team to face the administration's team. Therefore she, CFA President Terry Jones and the chapter president from San Francisco State, Tim Sampson, went down to Los Angeles to meet with Eliseo Medina, vice president of the SEIU.[8] (The CFA was now affiliated with the SEIU, as well as NEA and AAUP, and had been ever since 1983.) Fendel says: "We asked him [Medina]—we had a contract, it has been voted down, there is a lot of disarray going into bargaining, what should we do?"

Medina assigned them a consultant, Bob Muscat, who had been regional director for the SEIU, overseeing 52 locals. Muscat remembers how he felt about this assignment: "I was talking with my wife—should I take the CFA job? It's going to be easy, like semi-retirement. It's only one local, how hard could it be?[9] As it turned out, I probably worked harder and was more passionate about working for CFA than in the 25 years I worked for SEIU."[10]

Muscat came on as a consultant just before the election took place. He would later be hired by the new leadership as general manager of the union staff.

Things were now moving very quickly. Meisenhelder's next task was mounting the open opposition in the upcoming election of the five state-wide CFA executive officers and the Board of Directors.

WHY THE VOTE TOOK PLACE IN THE DA

At this point, we have to identify five official internal structures of the CFA to explain why this election would take place in the Delegates' Assembly, not with a direct membership vote. Each of the following statewide structures has some decision-making power.

First, there are five statewide Executive Officers. Second, there is a Board of Directors. These offices—the five Executive Offices and the members of the Board of Directors—are the positions to which Susan Meisenhelder's group hoped to elect their candidates. Third is the Delegates' Assembly, which consists of each campus chapter president plus at least one Lecturer from that campus, and other elected representatives based on membership. The total number of people who might participate in a DA meeting could be several hundred. The DA meets twice a year and elects the five statewide Executive Officers and the Board of Directors.

Fourth is the Presidents' Council (PC), made up of chapter Presidents. It meets more often than the DA and had some executive but mainly consultative power. Then there is the fifth structure—the Lecturers' Council—which has existed on paper ever since the collective bargaining election back in 1982 and was patterned after the UPC Lecturers' Committee. The Lecturers' Council was supposed to have a representative from each campus but often they did not really represent Lecturers; they were just a Lecturer who happened to be standing nearby when a chapter president was looking for someone to take to the meeting. John Hess called them "Hey you's" because the chapter president would call out, "Hey you! Are you busy next weekend?"

Meisenhelder and her group did not actually have a formal slate nor did they have explicit public positions that could be debated, aside from their opposition to the TA. They did not even have a one-paragraph statement saying what their goals were. They just trusted that within the DA everyone knew who was running with whom. Elizabeth Hoffman said: "People knew Susan. She was a chapter president. She'd been on the bargaining team and people knew that she'd been a Lecturer—she was public about that—and had gotten tenure. People on the Lecturers' Council could see what a big change it would be to have her as president."

It worked. All the old-guard officers were defeated and a new group of leaders was elected. Meisenhelder became president. This was the first time that there was an incumbent president who was not re-elected for a second term. Jane Kerlinger ran for executive vice president and also won. She says:

> You can imagine that people were saying there should not be two women leaders. Also that someone like me should not become vice president since that's the presumptive next step to be president of the union; a Lecturer had never occupied the VP position. But I had counted my votes.

Although this was the change in leadership that marked the beginning of the "Revolution," it was not the Revolution itself. That term refers to what happened in the years immediately afterwards, as the union transformed itself into an activist one, mounted credible strike threats, and successfully bargained the good contract we will look at later. By then, the CFA would have also become part of a national, even international contingent faculty movement.

There are several points to be made about this sequence of events and we will refer to them in later chapters. For now, we want to note Meisenhelder's comment about how hearing from "old leftists" in a moment of critical decision making was helpful. Another is the importance of organizing an opposition group that took pains to develop its strategy separately before becoming public. We will refer back to this later, when we talk about the Inside/Outside strategy. The opposition group organized itself first, so that its members would be accountable to each other and its goals. This group did not have a slate, but it had goals and arguments, held low-key meetings over four or five years, and built its strength partly on Lecturers who by now were the majority of faculty, if not the majority in any leadership positions. It demonstrated its strength in a showdown, getting the TA rejected. Only then did oppositionists run for election as top leaders of the union.

There were, however, consequences for the lack of a formal slate and program in that election. Meisenhelder and her group relied on their visibility and personal records within the DA to attract votes, but how this newly elected leadership would act as a team once they were in power was hard to predict. Elizabeth Hoffman said:

There were some people who were elected with her [Meisenhelder] who were not really on the train. She had officers who would actually more or less agree to something in an officers' meeting and then, in the large assembly, speak out against it. You know, that destroys a leader. She would have to run some people against them in the next election, and that was tough.

Reaction of the old guard

Jack Kurzweil describes some of the bitterness among the old guard after the election that brought in Meisenhelder's group:

When the dust settled, there were a whole bunch of people in the group that had helped in dislodging Terry Jones and electing Susan, but who still basically assumed the privileged position of being tenured white guys, including one Indian guy who was an honorary white guy. There was Susan up there bringing women into the leadership and these guys freaked out. The idea that there were women at the head of the union and some of them were Lecturers had these guys huffing and puffing around; not being able to use their stentorian voices was driving them fucking crazy. They started being abusive. Nominally, I was part of that crowd. I got stuck in the middle and didn't do enough to differentiate myself from them. In my viewpoint the cutting edge here was not anything about them not having run on a platform or having no explicit platform. It was about gender and privilege and old-white-male disease.

So even among the most vocal opponents of Susan's were some of the very same people who had worked to elect her. They supported her because she projected militancy and was the alternative to the old guard. But when she started to exert leadership and bring women and Lecturers into the leadership and insist that leaders actually do the work, many of these guys, including those who had supported her, just couldn't deal with it.

I tried to thread the needle because I had fought the good fight with these guys. I was wrong.

NEGOTIATING THE NEW 1999 CONTRACT

In support of bargaining for the 1999 contract, the new leadership quickly started trying to get the whole union involved and be much more activist and public. Time was of the essence. They hastily held media events on the various campuses and beyond. Meisenhelder called them "smoke and mirrors," but their effect was to create an illusion of widespread faculty support for the bargaining team and the possibility of disruption. It was not a real "credible threat of disruption," just borderline, but it was enough. The smoke and mirrors was mainly directed upwards, at the chancellor and the trustees. Nina Fendel says:

> The chancellor was a classic fat cat who did a lot of public speaking to higher ed conferences. We sent a bunch of people to these conferences. Our people got dressed up like for any academic conference. One time he was speaking at the Marriott in San Francisco and we decided to hold what we called the "Inside Game" in the hotel. We got amazing support from the conference staff—the hotel employees. We got floor plans, schedules. The conference had registration tables, but no other security. Our people had banners in their briefcases and waited until the right moment to unroll them in the back of the room where he was speaking. The banner said "LECTURER FAMILIES NEED HEALTH BENEFITS TOO." We did tableaus—Lecturers brought their children, to make the point. There was an article in the *Chronicle of Higher Education* and all it talked about was the union action.

Bob Muscat enjoyed describing these events:

> They went to every speech that the chancellor made, all over the country, and in the middle of his presentation in New York City or wherever, they would pop up out of the audience and talk about what it was really like. They would hand out leaflets and literature about what was going on in the CSU and why they shouldn't listen to that guy who was standing in the front of the room talking about educational policy.

Kerlinger saw the effect of the union actions at a bargaining session: "The chancellor walked into the room and took his jacket off and didn't even look for a table to put it on. He just dropped it on the floor. That was a sign that he was shaken. Susan and I were just giggling. We couldn't believe it. But they just wanted to get a contract. They just capitulated."

In fact, the contract they now negotiated was much better than the rejected Tentative Agreement. Meisenhelder attributes this achievement to a number of factors. One was that the administration was shocked at what had happened: "They were stunned. They could not believe the election didn't go like they were told it was going to go. So I think it was a very destabilizing thing."

At the start of negotiations, one of the old guard told Meisenhelder, "I like you, I think you're smart, but you're too naïve to be doing this right now." Later he called her back and said, referring to the rejected tentative agreement: "My hat's off to you guys, you proved us wrong; there was more to be got."

Meisenhelder also credits Bob Muscat. As a seasoned staffer from the SEIU who had bargained many contracts, he had years of experience in the rough-and-tumble of organizing and bargaining outside academia. Bringing this to the genteel professional culture of university faculty was more than just a breath of fresh air. According to Meisenhelder:

> After we got rid of the old crowd there were people thinking, we don't need another PhD on staff. What we need is somebody with some strategic chops and some labor experience. Even if Muscat wasn't a PhD he was really good at saying, "I really don't know anything about that but here's what I think from my thirty years in the labor movement." But he had no idea about education. He had never worked in higher education. So he was not that kind of help at the table. That was kind of smart. He made me do the talking at the table.

The chancellor, not surprisingly, assumed that the new top male staffer, Bob Muscat, was the brains behind the changes. When they were about to finish negotiations, this chancellor said to Muscat, "If we finish this contract, will you stop following me wherever I go?" Muscat recognized the chancellor's mistake:

> It wasn't me or other staff. It was faculty leadership that had the courage and dedication to do that. They made the choice, they had the courage to stand up, they knew what to say, they got thrown out of those conference rooms—having people at the top of the union who were willing to do that was what changed things in the CSU and brought about this contract.

Muscat described Meisenhelder's manner at the table:

> She came in and did an opening statement—just like we had talked about—defining for them what this was all about. She was a brand new president taking over with a brand new style, she had to define for everybody at the table that she had become the new president, what was in her head, what was important to her. She did as good a job as I could have done under the circumstances. It was stunning to me, it was attractive, all the leadership potential that they had—so, given that I was planning to retire from SEIU, I started talking to them about the possibility of coming on at CFA as general manager.

They changed his position from consultant to general manager, and he got to work right away. Muscat said:

> In terms of the Lecturers, it is like a dream for somebody like me to have access to members with the capacity for hard work and skills that the faculty had. We definitely started from the very simplest things—like going to their executive board meeting and seeing faculty members drift out and come back fifteen or twenty minutes later and fully expect the conversation to be where it was when they left, and that they were entitled to keep talking when they returned. I even bought Susan a gavel to use to make people stick to the agenda—all those fundamentals of running a good organization; it didn't take me an hour to convince her of this, it took two minutes.
>
> We had to renegotiate the relationship with some of the staff. They had a guy who had been their negotiator, a more traditional negotiator—we had to put him to work, really, because he was getting paid without working. There was the ex-general manager, we had to send him on his way. We had to get the lobbyist to retire—he was making so much money—he would walk in and say, "I made $15,000 this morning I don't even know why I come to work." Well, he wasn't really coming to work.

Elizabeth Hoffman expanded on the problem with the old staff:

> It's more work for staff to do the organizing model. For example, a huge sea-change was tracking. Staff always said that a lot of people were

coming to meetings. But how many, who? The new CFA—you have to have names. It was hard—you have to get people to sign, like if there was going to be a strike vote. You could only count the people who signed. Some staff would say, "I talked to a lot of people," but we couldn't count that. Still there were staff people who would say 85 people would turn out, but then only 15 would show up. We had people fill out a card and we called and reminded them. The big meeting where we got 1500 people was when everything was tracked. This was the first time ever, it was so different. A lot of the older time leadership really resented that. Some of the staff tried to undermine it. They would go back and work in league with their alliances with governance and staff.

Muscat described a backlog of arbitrations: "They had about 350 cases that were waiting for arbitration. That means you were fired, you'd arbitrate the case, it would take three years, that throws the whole idea of arbitration out the window. You'll die, people move on!"

And lot to learn about public relations:

They showed me a *New York Times* ad that they ran. It was all text, no white space, no sub-heads, nothing—and they apparently negotiated among themselves for two months before they got the language that they wanted! I had to teach them simple things, like having artwork. I said, "You are all academics trained to sit back and fold your arms and pull apart plans and say what's wrong. That will never work. You'll never get what you want that way."

He had a diagram he would show people:

I had a little chart that I used. I used it so much that when I retired at my going away party they gave me a T-shirt with the diagram on it. It's basically a triangle over time: leverage builds.[11] You have to put your plan together to build leverage. It's embarrassingly simple. If you are going into negotiations, you decide what is the most advantageous time for you to settle. You wouldn't pick a time when the trustees are on summer break, for example. The moment you want to settle will never hit at the time that you want it to hit if you don't know what time you want it to hit. You have to pick a time and then you put your plan together to make it consistently worse for the other side as time goes on, right up to when you want it to hit. So you don't bring up your salaries here—where the leverage is low—you bring your salaries up here—where your leverage is close to the maximum. Then you sequence your bargaining according to the plan that you have. If you have to adjust your plan, it's not the end of the world, but you'll be in general confusion if you don't have a plan. You do with this a calendar where you're breaking them up into communications, political action, fieldwork, different categories of events and activities to build that power to the deadline.

If you settle too early, you realize that your power would have continued to increase. If you go on longer and settle too late, you'll see that your power will have begun to decline. So you need to hit it right. You have to build pressure all the time, maybe even for a year, so that the decisions you make late are much tougher than the ones you make early.

In that interview, done in 2016, a decade after Muscat's retirement, he was still delighted with his work with the Lecturers:

These Lecturers were people with a different kind of energy. Honestly, it's not as if they're not as bright or capable as the tenure-track faculty. But there was a working class style to them—all of them; it was kind of interesting, there was a hardness or determination in them that was different from the tenure-track faculty, that gave them the resourcefulness and capacity to do things for themselves that didn't take a lot of convincing on my part.

Bob Muscat looked back on his final effort for the CFA before he retired (for the second time!). It was the building of one long, big campaign that built a credible strike threat that forced a settlement:

The last thing I did for them was a system-wide rolling strike. I think they settled a day or two before the strikes were supposed to start. I was the first one to get them ready to strike, to turn CFA into a union that could strike, which would only have been possible by building up the Lecturers in the union. I saw CFA as really one big campaign. The nature of the people who are attracted to an organization changes, the organization changes. When we settled, we had a press conference with fifty people in the room and some reporter asked the president of the union, was there anything you wanted that you didn't get? And he couldn't think of a single thing. It was kind of the high point of my career. This was just before I left, my last contract, after eight or nine years.

Readers might wince to hear Muscat say "I did *for* them." But in fact, in our interviews, all the people who were active in the CFA at this time said "I did this, I did that," taking for themselves a personal share of credit for the union's successes.

4
"They have nothing to teach us"

In 2000, the California legislature passed the Fair Share Act, which mandated that all public employees in California who were covered by a union contract should be charged the "cost of representation." Everyone in an exclusive bargaining unit in the public sector, in other words, would have to pay what was called their "fair share," usually an amount slightly less than dues. In some states, this payment is called an "agency fee," because it is a fee for the union's work as the bargaining *agent*.

The money that rolled in from the Fair Share Act changed a lot of things for the union. For example, it was enough money to provide stipends and travel expenses for Lecturers who were assigned to do organizing on their own campuses.

Some things that worked in favor of organizing in the CSU system happened because of legislation at the state level. While local collective bargaining and contract enforcement may be what come to mind first when developing a union strategy, community organizing, internal organizing and political action should all fit together. We talk about this when we talk about the different terrains of struggle. This is the organizing model that Elizabeth Hoffman, who became academic vice president for Lecturers, refers to as "the Octopus," i.e., a creature with many arms. She would draw a cartoon graphic. "What's the whole picture?" she would say. "The idea is that you have to show that you do everything simultaneously. Working with other allies—students, everybody—it's not just strategic, it's practical. It can be reassuring, too, because not everyone has to do everything. There was hardly ever a meeting where we did not put up the Octopus."

Critical events at the state level go back to the law enabling collective bargaining for employees in higher ed, HEERA, enacted in 1979. Then, in 1994, Assembly Bill 211 passed, providing funds to be used for access to healthcare for Lecturers working 50 percent of load (later expanded to include those working 40 percent of load). Then came the legislation that brought

Figure 4.1 The "octopus" symbol used by CFA activists to symbolize the importance of "doing it all at once." A stuffed toy octopus was always on the table at bargaining sessions.

Fair Share revenue into the union. Readers who live in states where legislative action on behalf of working conditions for higher ed employees is particularly unlikely may feel frustration and think, "That can't happen here—I thought we were talking about bargaining, not politics!" Our hope, which we will talk about more in the section on what we as contingents fight for, is to make clear that this fight takes place on all the different terrains where power matters. Sometimes, the potential for building power is limited to the local, campus, or even departmental level. Sometimes it is on the individual level and we can build it up from there. In others, it is possible to move directly to a larger scale. Either way, remember that we are describing a transition in a union that took place starting in the late 1970s, nearly twenty years before a national or international network of contingent faculty began to knit together.

The logical argument for a "fair share" fee is that everyone, whether or not they have actually made the positive action of joining the union and becoming a member, is still represented by that union. A union has the legal responsibility to represent them properly in both bargaining and grievances, without discriminating against them on the basis of their not having joined. It is not unusual for a nonmember (sometimes called a "fee-payer") to suffer some kind of injustice at work and come to the union for help. Legally, the union then has to devote time and resources to defending them because in fact they have no other place to go. There is no alternative union; their union is the "exclusive bargaining agent" for that person. Many unions through this process manage to demonstrate the value of the union's support to fee-payers, and they voluntarily become a member.

Getting Fair Share through state legislation, as happened in California in 2000, is unusual. In other states, a Fair Share practice is generally something that is bargained, not gained by state law. In fact, Fair Share in the public sector all over the US was nullified by the US Supreme Court's *Janus* decision in June 2018.[1] But at the time, the Fair Share Act in California directed a large amount of money to education unions.

Once the law mandating Fair Share was passed, the CFA decided to organize the faculty to hold a vote to implement it. Since the fees would come out of people's paychecks, this required some effort.

The San Francisco State chapter president and one of the Statewide VPs at that time was Tim Sampson. He was a nationally recognized organizer, an Alinsky-ite.[2] He had taught at the Highlander Center in Tennessee and was part of the Center For Third World Organizing, as well as a founder and leader of the Association of Community Organizations for Reform Now!, or ACORN, in California. He had mentored hundreds of people (including Nina Fendel and John Hess, as well as Joe Berry). Sampson decided to make it his signature task to bring the fee-payers into the union as members. He said that his best argument, aside from the way the money could be useful to strengthen the union, was that the amount of the fee was about 80 percent of the cost of actual membership. The remaining 20 percent would bring with it the right to vote on union issues. He gathered the necessary paperwork about the agency fee issue and wrote individual emails to every faculty member who had complained about having to pay fees. He persuaded many hundreds of faculty to join as members.

Ultimately, the members voted to implement fair share. With it came a great increase in voluntary full membership. This brought an entirely new character to the union finances.

JOHN COMES BACK TO SF STATE:
THE STUDENTS ARE DIFFERENT

John's first impression when he came back to teach at San Francisco State in 1999 was that the overall culture of the university had changed. The old San Francisco State of the 1968 strike days and the Movement was gone from the consciousness of his students, if not from the older faculty. In its place was a defensive, anxious, fearful attitude that had developed during the financial crises of the 1990s. As one of the founders of the academic field of film criticism, John's teaching goal had been to instill a critical, questioning perspective into the minds of his students. Now he found that this kind of political angle frightened people. The chill had sunk in deep. This was partly due to a change in the student body. They were just eager to get into the business and make a living. The years of "greed is good" had made their mark.[3] John said:

> It seemed to me that in the 1970s, 1980s, and even early 1990s the students had been eager just to learn. Even though they may have been handicapped educationally, they would trust you enough to dare doing whatever you asked because you were the teacher. When I got back, people didn't seem to have that intellectual curiosity. They didn't want to think about what things meant. The emphasis was all on production. They just wanted to make movies. Film criticism was just something they had to do in order to graduate. They would say, "I want to go to Hollywood!"

John was right about the demographic change; there was a reason for it. One aspect was tuition: when he first started at SF State, there had been no tuition fee. By the late 1990s, students were paying tuition, which kept rising. They overall were older and had now taken on debt or were working, often full-

52

time, while taking classes. To him, this focused them on beating the odds rather than doing something political that would change those odds.

Nina Fendel remembers what John was like when he first came back to SF State:

> I saw him in the faculty club at SF State where people would go for lunch. I invited him to something and had a one-on-one with him to explore what might he be interested in, in terms of becoming active. He was hesitant. He expressed that he was disappointed that nothing had been done to save his job back in 1993. He didn't want to put a lot of energy back into the union. I invited him to some upcoming union event. He was reluctant. He didn't want to step into any role with the union.

We should also remember that he and Gail had just relocated because of one of the most depressing turns of events in contemporary labor history, the dismantling of the Carey reform leadership of the Teamsters by the cynical Hoffa old guard. Nina said, "I respected that and backed off. But lo and behold, he showed up at the event. I tried not to press him beyond his comfort level. But then he started to show up. Everyone was glad to see him and he had a lot of friends. He started coming to union things, but he didn't assume an official position."

Organizers will recognize Fendel's persistence and the numerous "touches" required to encourage even this experienced activist to come back to the struggle. The came the Fair Share election:

> We won the Fair Share election and we signed up a whole lot of members statewide and our budget doubled. We needed staff. Our seven reps had four campuses each and some of these are really far apart. You can't do more than put out fires and deal with emergencies. Plus, the reps did most of the arbitrations, so they were really working hard. We only kicked arbitrations to lawyers if they were very tricky or sensitive, or if we thought people would sue. I was the only lawyer rep. So all of a sudden, we could increase our staff.

Despite his original reluctance to getting involved in the union again, upon Nina's suggestion, John applied for and got a position as a staff rep.

THE LECTURERS' COUNCIL: A BACKWATER WITHOUT LEADERSHIP

John scoped out his new position by going to see what had been happening with the Lecturers' Council. The current chair was a man from Chico. It was a depressing sight. John said:

> I flew down to LA and went to one of these hotels and walked into the room where the Lecturers' Council was meeting, and I blurted out: They're all so old!! They were the same people who were there when I left seven years ago. The president of the Council didn't have an agenda for the meeting. He would just ask a philosophical question and people would answer it. There was not anyone in the room who was willing or able to disagree with him. It was all just pissing and moaning. It was an empty thing.

Elizabeth Hoffman, who was on the board of directors as a Lecturer representative, called the Lecturers' Council "lackluster." Participation in the Lecturers' Council in 1999 had sunk so low that not every campus reliably even sent a delegate.[4] Elizabeth said of the new chair, who had replaced Jane Kerlinger:

> He was a disaster. A lot of people up at Chico loved him, he wasn't a bad person, he was in a lot of progressive left-wing sort of stuff, I think he started Burning Man, but he was completely disorganized, never was on time for anything, everything was chaotic.[5] Eventually, it so happened, we let him be on the bargaining team—except that he was terrible—there was a point at which he went and had a private talk with someone in the chancellor's office and then he was never allowed on the bargaining team again after that.

The "empty thing" that John saw was a gathering of demoralized people. First, they had been beaten down by the bad contracts of the mid-1990s. Then the election for the executive board and the board of directors that Susan Meisenhelder and the other new leaders won had taken place in the Delegates Assembly, and Lecturers did not participate much in the Delegates

Assemblies. So the drama at the Delegates Assembly passed over the heads of most Lecturers. As far as they knew, no positive change had taken place.

But after the election and during the subsequent intense bargaining, there had also been little or no mobilizing on the campuses except for the events that Susan Meisenhelder referred to as "smoke and mirrors." These were intended to display faculty support but were directed mainly upwards toward the trustees, not out horizontally to the faculty. They did not represent real organizing. The "smoke and mirrors" did not reach down into the Lecturers' Council.

The strategic planning session that led to the White Paper

But while the organizing among Lecturers had lagged, the urgent need for clarity about the role of the union and the future of the university was stirring up turmoil at the top level of the union. Many of the old guard (mostly men) had been taken by surprise by the success of the Revolution, which had sent the message of rejecting the service-oriented bureaucratic culture of the CFA. Now the Fair Share election was bringing in dues money, as well as fees that would support the work necessary for implementing change. The reaction of some of the old guard was to argue that the union had a "moral duty" to spend the new money on lowering the dues. Bob Muscat had been hired as general manager and some of the old staff had been sent packing, which also upset old loyalties. Susan Meisenhelder and Jane Kerlinger were both women in positions traditionally held by men, and many of those men still held leadership positions in the various chapters and were disaffected if not angry. There were many deep tensions throughout the union, tensions that were increasing.

Tim Sampson approached Katie Quan, a labor educator who was at the University of California Center for Labor Education and Research in Berkeley. She had been the elected vice president of the Union of Needletrades and Industrial Textile Employees (UNITE) and had herself worked in a sewing shop for many years, before being hired at Berkeley. Sampson proposed that she lead a strategic planning session for the board of directors of the CFA around a vision that would be activist oriented.

Katie Quan interviewed Susan Meisenhelder and about five of the other board members. "They all had personalities," she says, smiling. She decided to give it a try. Here is her description of that meeting:

It took place in Southern California, in a hotel. We were seated around a table, not in rows: it was a hollow square. It was a little inflexible, not good from a labor education point of view because when people are talking they are far from each other and they can't really see each other's eyeballs. People had to raise their hands so it was hard to have a spontaneous discussion. But I decided not to move the tables because it was the best arrangement for hearing people. I could see that there was in fact division between the newly elected people and the old guard—people who weren't actually old in terms of age but who didn't share the activist bent—that is, the group was fractured, and I wanted everyone to hear everything. It had to be a big room because it was a big group, but it had to be all one discussion. I placed myself in the middle of one of the long sides. Behind me was one of those fold-out walls. I put flip chart paper up on the walls behind me. I had to use tape because it wasn't sticky.

I had decided to do an evaluation of the union's strengths, weaknesses, opportunities and threats.[6] I said that the way we would utilize this evaluation was to think about how we seize our opportunities to build on our strengths which then overcome our weaknesses and neutralize threats. I put it all into one sentence.

I set some ground rules, that everyone was equal. That may have given some indication to the non-activists that I was serious about being a non-partisan facilitator. I tried to make sure that the non-activist people felt their voices were heard and that the activists did not overstep boundaries.

I suggested the categories—strengths, weaknesses, opportunities and threats—and they did one category at a time. Then they broke for lunch. While they were at lunch, I organized what they had written.

Nina Fendel was at this meeting and said:

Katie did a really excellent meeting. She asked people for ideas and had them write their responses on post-card size post-its. Then we had a break. While we were taking a break, she organized the post-its. We had an hour for lunch so she had an hour while we were out of the room. When we got back, she fed them back to us in an organized way. It was really wonderful. Faculty are used to having a lot of air time and are so long-winded. This was thirty or forty people. I was in awe.

Katie described what they did after the lunch break: "When they came back in I physically moved the flip charts. I picked them up and carried them, moved opportunities into the first place, so strengths were next, to overcome weaknesses, and so on, so they could visually see the progression." This was the physical expression of "putting it all into one sentence." Katie described the reaction:

There was this one woman—I believe she was from CSU Long Beach— who had been identified to me as part of the non-Susan group. She spoke up right after I moved the flip charts. She didn't raise her hand. She just said—no one else was talking—she said, "What we could do is ... ." You could tell she was thinking while she was talking. You could just see the bing-bing-bing-bing light bulbs go on when she said it; it was pretty remarkable. We were in this situation where people were thinking in a non-partisan space. So she just blurted it out. There was immediate unity around it. I think everyone thought it was a brilliant idea. It was not an agenda item, it was not a proposal on the table. It was a new idea that quite definitely got hatched at the table. If Susan had suggested it, it might have gone nowhere.

Her idea was to prepare something for public consumption to counter the administration's perspective on the CSU system, setting out the situation, from the union's point of view, for other faculty as well as general readers who were concerned about California higher education.

This woman didn't use the words "alternative," "counter," etc. But what came out was the recognition that the strength of the CFA was that it could develop a progressive vision for the whole CSU system that was in contrast with the chancellor's vision, and that this would be the guiding document for the union and it would serve as a platform around which the union could mobilize its membership. You might think this was not such a new idea, but in fact nobody in the room had thought about it and nobody outside the room had suggested it either.

Having a White Paper was something they could all agree on. If it had been "Shall we call for a general strike?", you wouldn't have gotten that kind of unanimity. But everyone knew there was a need for a unified vision. They also had the people and the wherewithal to do this. I come from the garment workers union and it would have been difficult to feel that my E Board could have come up with this solution to this problem.

But they were from a union of teachers, academics, researchers, who were both familiar with educational problems and the needs of both students and the university, and they were also savvy about education and educational policy, so this was the appropriate solution for this group.

Looking back on this in 2019, Katie said:

I count this as one of the most successful strategic planning sessions I have led because of the conclusions that we reached in the meeting as well as the implementation afterwards. It's been my experience that a lot of unions do strategy planning but are unable or unwilling to carry out the results. In this case, they were able to do that.

AN EFFORT TO ENVISION THE FUTURE OF THE UNIVERSITY

One aspect of carrying out the results of that strategic planning session was to hold a conference to be called "The Future of the University." This conference was to open at San Jose State and then circulate around all 23 campuses. Jack Kurzweil, who was still president of the CFA chapter at San Jose State at that time, was planning it. He envisioned it as a way to embed collective bargaining goals into the framework of the future political economy of higher education, in order to push the union to be the focus of a social movement around higher education. The conference was in support of the development of the White Paper, which was explicitly supposed to be an alternative to a document put out by the CSU chancellor setting out his business model vision for the university.

John Hess, by now on staff for the CFA, was still on his first assignment to do organizing work at SF State and at San Jose State. Later, this would be changed to working only with San Jose State and adding the Lecturers' Council. At San Jose State, he met Jack Kurzweil, who took him around and introduced him. "That gave me very good PR," John said. But providing staff support to the conference put John in the middle between Jack Kurzweil and Jeff Lustig, another strong leftist union leader and the chapter chair at Sacramento State.[7] John remembered this with a laugh:

Jack really got into it. Him and this guy Lustig were the two guys working on the conference. It was one clown show short of a circus! Here's these two Jewish leftist activists, different lines of activists, different types of Stalinists—and me, my job was to keep these two guys from killing each other. They called each other all kinds of names and wanted to be in charge. But of course, it was Jack's campus.

John was learning the difference between being union staff and being a colleague. There really wasn't six degrees—or even two—of separation politically between Jack and Jeff but that didn't stop them from arguing. In fact, Jeff and Jack lived around the corner from each other in Berkeley and had known each other since Free Speech and Anti-War days at Berkeley. Kurzweil said: "I was a Communist and an electrical engineer, what business did I have having political opinions? Compared with Lustig, an SDS guy who had a PhD in Political Science and had started the Center for California Studies? That was the kind of dynamic it was." "Clown show" or not, "old leftists" like Kurzweil and Lustig would bring needed perspectives to the union at key moments.

Ultimately, the conference took place at three campuses. It was considered successful, but John, observing it, realized that something very important was missing: "It became clear to me that the conference talked about things as if they were still coming in the future, things that in fact had already passed. For example: the corporatization of the university. The university was already corporatized. They had won! We weren't going to up-end that."

Corporatization included the wave of new hires into non-tenure-line Lecturer positions, something John had personal experience of. He felt the urgency of the situation but was worried that he had not played an effective role in the "Future of the University" conference. He was also depressed by the "empty thing" he had seen at the Lecturers' Council. He turned to Nina Fendel again. He needed a colleague. Governance, or the practice of leadership and exercise of power within an organization, is a tricky thing, especially when you have elected leaders on the one hand and paid staff on the other. He had now moved from being an activist member to being staff. Where could he go to get some advice? Nina told him to go up to Butte County, north of San Francisco and get together with Jane Kerlinger who was by now executive vice president of the CFA.

John went up to Chico, found Jane's house and met her. Jane remembers:

John stayed with us. We had dinner and talked. He sounded tentative and didn't seem to know where he should fit in regarding governance—what does staff do? Do I just support the elected leaders, or do I actually run the thing and see how far I can go and get away with it? What should he do about the Lecturers' Council? John felt like he might make a huge blunder. From having been Chair of the Lecturers' Council, I knew what needed to be done. You had to have conference calls, put out an agenda, all that. Since I was not a Lecturer any more I didn't go to the Lecturers' Council, but I understood what was going on.

John was just coming to realize what a staff person could do. The elected people are supposed to be running the organization, but in reality, they may not have the training, skills, time, or interest to do the job. Jane elaborated:

There are certain aspects of the governance that you have to work with. Other parts, you take their strong points and let them run with it. Then everything else that has to be done, you do it. One chapter president might not like to make plans for what day we're going to do this and what day we're going to do that. So you have a way to make him make a commitment. You make it look like they're doing the work.

It happened that the current president of the Lecturers' Council, the person John had observed in Los Angeles at the Lecturers' Council meeting where there was no agenda and people spent the meeting raising questions that did not lead to answers, also lived in Chico. He found out that John was at Jane's house and decided to drop in. At dinner with John and Jane, he dominated their discussion with his long-winded and depressing opinions. After he left, it became clear that John and Jane were equally frustrated and angry. However, because of their long conversation that afternoon, John had much more confidence that it was possible to take control of the situation.

Between the two of them, they laid out a structure for organizing. John sat down and typed up a draft of it on Jane's computer. "It wasn't rocket science," said Jane. These things had all been done before, many of them when she had been the chair of the Lecturers' Council. But now there was the Fair Share money to spend, and new top level leadership, which included Jane. Between the two of them they laid it out all in one document.

The plan included chapter Lecturer meetings around the state on each campus, all within the span of a single week. There would also be a monthly phone conference call for all the Lecturers' Council members and others who wanted to be involved. These would be open to anyone who wanted to join in. The idea was to get people excited. Then, since there already were two Delegates Assembly meetings a year, fall and spring, and Lecturers' Council meetings which were supposed to coincide with that the DA's, they built on that by adding two more Lecturers' Council meetings, one in the early fall and one in January, making four annual LC meetings. These would be weekend meetings. On Friday night, they would have a speaker and on Saturday they would organize. With the revenue from Fair Share coming in, there was enough money to pay for this—for travel, for hotel rooms and meals for participants.

Before the next Lecturers' Council meeting, John and Jane talked with Nina again and she suggested that, since they anticipated that the Lecturers' Council president might show up without any agenda, they could write up their own in big letters on a big sheet of butcher paper and hang it on the wall in the front. Then when people came into the meeting in the morning, it would be there. In fact, that is what they did. John remembered that when the current president came in, he looked at it and shrugged and said, "I guess that's OK." And that became the agenda for the meeting.

In his interviews, John told this story two or three times, always with pleasure. That action was a marker that gave him the confidence that he could really assume the emerging role that he visualized: to be staff organizer for the Lecturers. Commandeering the agenda of that meeting was perhaps technically an interference with democracy, but no one raised a challenge to it.

Soon thereafter, John's job changed to make him half responsible for the Lecturers' Council and the other half staff rep at San Jose State. Jane Kerlinger was offered a similar staff job in January 2001. She left her vice president post and became the rep for the Sonoma State, Sacramento and Maritime (Vallejo) CFA chapters. These were good, full-time CFA staff jobs with benefits, retirement and a staff union. John and Jane were the first staff hired from the bargaining unit, to represent people with experiences like their own, something that the CFA had not done in the past.

THE LECTURERS' COUNCIL TAKES OFF

With new energy around, Lecturers started getting involved. Slowly but surely, the people who had accidentally stumbled into the room were replaced by people who deliberately sought it out and were interested in doing something. "Something" meant organizing. They used their meeting time to teach organizing, using staff people, labor educators, all sorts of different folks. Looking back on it, John says:

> Teaching the organizing was probably the smartest thing we did. These meetings were incredible. People were hungry. Some of them were like revival meetings. They were very energetic—a lot of energy. We had to keep bringing more chairs into the room. All the guests—AAUP, SEIU— they all came to it.

John acknowledged that the money made an enormous difference in resources that could be devoted to organizing. The CSU workforce of the 23 campuses is dispersed from the top to the bottom of a state that is the same size of many nations. This means that the costs of travel to bring people together can be prohibitive. With the Fair Share money, meetings could be held in hotels and people could get their travel expenses reimbursed. The Lecturers' Council organizing trainings also drew in the leaders of the union. John said:

> Well into this process—several years in—I realized that Susan Meisenhelder and Bob Muscat, the two lead people in the union leadership, the president and the general manager, came to all the Lecturer Council meetings, sat and said nothing unless somebody asked them a question or asked them for their views. That was an amazing thing to realize. Ultimately someone asked Susan why she came to the LC meetings. She said, "Because they're so hopeful. They are the most interesting meetings we have." I take credit for getting this started, but they picked up on it because it was something they wanted.

The tenure-line faculty began to show up too. Chapter presidents even began to come to these meetings—people who, John said, had for many years resisted any kind of training. The idea that after years of being on the defensive, something positive was going on, was a tremendous relief. At some point, even the chapter presidents suddenly started saying, 'We want to learn organizing. We want to be like the Lecturers' Council!' They began to get organized:

So we basically switched over so that rather than have the Lecturers' Council meet and do organizing, and the Presidents' Council meet and do organizing, and the Women's Caucus and the People of Color Caucus, the Third World Caucus—they all said "We want to do organizing!"— and the union was very happy to accommodate them. We switched over so that we would meet together. We would have these mass meetings. Particularly at the beginning of the year, summer and fall, in a big hotel in downtown LA, teaching, practicing organizing. It was like this full-tilt effort to bring the faculty into this and teach them organizing. Things really changed. Things changed, you could see the change.

It went pretty quickly. It was three or four years, that doesn't sound quick but it is, in terms of how these things are measured.

The purpose of all this organizing was to prepare for upcoming bargaining. There would be two important bargaining cycles in the 2000s. Preparation for these two bargaining sessions was the work that Bob Muscat, looking back on his career in the CFA, pointed to as his proudest moments: the CFA would have to be ready to mount a rolling strike across the entire system, a credible strike threat. A credible strike threat meant one that would be believed not just by members but also by management as well as the general public, including the students. Projecting a credible strike threat required preparing for a mounting sequence of contingencies and it meant using tactics up to and including direct action.

DIRECT ACTION, THE RUCKUS SOCIETY AND THE END OF "THE TENURED GAZE"

The Ruckus Society first became widely known for its work during the 1999 World Trade Organization (WTO) demonstrations in Seattle. They do trainings for demonstrators on making the most out of non-violent direct action, making it newsworthy, theatrical, photogenic, memorable, vivid, fun and at the same time, effective.[8] John said: "We were not going to beat these people without direct action. I had always heard about the Ruckus Society so I contacted them—a few of their guys who worked out of their office in Berkeley—and asked, 'What do you do? What is it like? Would you come and teach a session on organizing and direct action?'"

Historically, the education unions like AFT, AAUP and NEA were not open to viewing a group like the Ruckus Society as having anything to teach faculty. That would be "unprofessional." The CFA had already been criticized for doing some things that were called "unprofessional" such as paying students with Fair Share money to organize students to support faculty. But ideas like these came not from the top leadership of the union but from the bottom up, from people who had experience in the broader labor movement and social movements, especially the faculty of color, who were under-represented as active members of the union but for obvious reasons, close to the increasingly numerous students of color.

Joe Berry, who by 2000 had left Iowa to go work in Philadelphia as an organizer for nurses, and then moved to do Labor Education with the University of Illinois in Chicago, was invited to come out and speak at the same Lecturers' Council meeting at which the Ruckus Society made their presentation. He stayed to participate. There were about fifty people in the room. The Ruckus Society people took the first ten minutes to orient the Lecturers to the idea of direct action, to the idea that putting real, material pressure on your opponent was an essential aspect of any struggle of substance. Then they broke the participants up into teams and created a situation that wasn't exactly the CFA's situation but was close enough. Within each team, a role-play was acted out that gave each person a chance to say what they were willing to do and what were the limits of what they would do. Would they sign a petition? Would they march across campus holding a picket sign? Would they block a door? Would they occupy a building? This was with the understanding that each of these actions could have consequences for the struggle and for them individually. The exercise put the collective strategic and tactical decision making directly into these groups in a way that modeled grassroots democratic process.

The ability to implement these actions is the pattern that Jane McAlevey calls "structure tests," the testing of the union's capacity to engage in higher and higher level actions against the boss. They are also similar to what Bob Muscat described when explaining his approach to preparing for a strike.

After long discussions about their willingness and their limits, the participants came back together in a big group. People then role-played direct action activities, including physical linking of arms, sitting and lying down, facing physical opposition from the police or others, dealing with people who got isolated in a crowd situation, or with someone who was engaging in provocative behavior. At every turn, the goal was to learn how to de-escalate the confrontation, keeping it non-violent while escalating the campaign, pushing it forward. It was essential to maintain the collective unity of the group, physically and politically, and keep people in a thinking mode.

This exercise highlighted above all that physical collective action did not have to be mob action. It could be thoughtful; it was not beneath the dignity of faculty, it was fun and created a kind of joyous solidarity in that room. Everybody in the room realized this.

John Hess, watching the effect that the presentation was having on the Lecturers, and remembering how tenure-line faculty had been quietly showing up at the Lecturers' meetings, had a sudden flash of insight: an important relationship had been upended. Tenure-line faculty were supposed to be the ones who would instruct Lecturers about teaching, research, their students, all aspects of life in academia. Tenure-line faculty were supposed to be the ones who had figured out the secret of getting hired and promoted. In this relationship, Lecturers were the children and the tenure-line faculty were the adults, if not explicitly, then at least implicitly. But not now, not in the context of a fight for decent working conditions for faculty.

In one of his interviews, John looked back on this moment and recalled the moment when the penny dropped: "Suddenly I realized, they have nothing to teach us. They—the full-time tenured faculty—have nothing to teach us. We can teach *them.*"

For John, this was a seismic shift. He after all had been a Lecturer himself. The Lecturers, not the tenure-line faculty, possessed the critical knowledge that was required to move forward on behalf of faculty and students in the rapidly shifting world of the corporatized university. This was not just an academic or epistemological shift. This was an emotional shift, a shift in the scope of a worldview of what Lecturers collectively could be and accomplish. Lecturers were trained by the culture of academia to think that the tenure-line faculty possessed some secret about how the university worked, and that this explained their relative success compared to contingents. But this didn't have to be the way things were: Lecturers could suddenly come to realize, "No, they don't. In fact, we do—we know what it means to fight, and we can teach them."

Suddenly there was an alternative to "the tenured gaze," the view from the seat of privilege. The alternative was collective organization for power.

Bob Muscat had recognized this early in his work with the CFA. It got him excited about Lecturers. He saw it "as hardness or determination in them that was different from the tenure-track faculty, that gave them the resource-fulness and capacity to do things for themselves."

Years later, Jack Kurzweil remembered talking to John during this period. Jack couldn't quite believe what he was hearing:

> I asked, "John, do you really think you can get this faculty to go on strike?" John said, "Well, we've got this guy from Ruckus Society, he's teaching us this." I said, "But do you really think they'll hit the bricks? I remember how much it took to get two hundred faculty at San Jose State to go out for two hours back in the '90s." John said, "You don't understand, it's different. It's different." And he was right. I was stuck in an entirely different model, an old model, of how you do things.

The lessons and practice of direct action laid the basis for building the credible strike threat, the system-wide threat that really did have the capacity to bring all 23 campuses to a standstill.

PART II

HIGHER ED WAS NEVER A LEVEL TERRAIN OF STRUGGLE

5
Four Transitions and How Casualization Served Managers

We now turn to a brief history of higher education in the US, focusing on the workforce and the conditions under which faculty worked. Most histories of higher education treat it as a single story of increasing expansion, specialization and democratization, viewed from the top down. We, on the other hand, periodize this history by the surges of conflict that have created key transitions, looking at them primarily from the point of view of students and faculty, bottom-up rather than top-down. Keeping in mind that "our working conditions are our students' learning conditions," we focus on the workforce. As always, we ask the fundamental questions: for whom, by whom and for what purpose?

It is clear that education, including higher education, is about to undergo a major transition. However, the current transition is not the first. There have already been four. In each case, the institution has kept the same name—college, university—and under that name it has engaged in the same fundamental purpose: to gather, evaluate, protect, create and pass on knowledge of the present and the past, and to provide a structure for the people who do this work. Each transition has had significance for the higher ed workforce in terms of demographics, working conditions and role in society. Each transition has provided a different answer to the three basic questions: for whom, by whom and for what purpose? For whom was higher education intended? By whom was it carried out? What was its specific purpose at that moment? As the moment of each transition approached, conflicting answers to those questions were fought out on various terrains, just like today. Then, under pressure from below and above, the whole system, or institution, adapts in a way that means changes both deep within and on the surface.

None of these should be thought of as "progress" in the sense of things always getting better, nor should "reform" automatically mean improvement. Each transition is actually a re-set following a struggle, a temporary and reversible truce between two or more forces working out the social and historical dialectic. This is what we are in the middle of today.

This chapter, therefore, looks back at the four great transitions of the past. We call these standardization, expansion, the Movement, and neo-liberal reform. Casualization—also called contingency and precarity—is part of neo-liberal reform. The previous transitions that have re-invented higher education can help us imagine how profound the changes ahead might be.

STANDARDIZATION

By the late nineteenth century, the US had undergone a very rapid industrialization process and the two new major classes—industrial capitalists and industrial workers—had emerged. It is during this period that the United States became a mostly urban country and a new segment of the middle class appeared, made up of corporate managers, accountants and communications experts.

The new industrialists saw the need for better mass education for the working class, as well as higher education for their assistants and managers of the huge new enterprises in manufacturing, transportation and finance and, later, in the public sector. These needs created a demand for more schools and therefore more teachers at the college level. But at that point the actual work of these teachers varied enormously from state to state in form, character, purpose and quality. A degree from one institution might mean something very different from a degree from another institution.

The first great transition, therefore, was the standardization of teaching work, an idea promoted by industrialist Andrew Carnegie, the author of *The Gospel of Wealth*, and the same man who founded the company that would become US Steel. Carnegie is also remembered as the man who oversaw, from the distance of his hunting lodge in Scotland, the violent crushing of the Homestead Steel Strike in 1892.

In 1905, he set up the Carnegie Foundation for the Advancement of Teaching, which commissioned a report that laid the basis for the future of education in the United States. Dealing with both K–12 and higher ed, the report recommended that college classes be structured on the basis of a "Carnegie credit unit." The typical class would be three credits. It assumed that one credit would represent one hour of classroom lecture and two hours of study outside of class each week, and spread over a 15-week semester. The familiar formula in use today is built on this concept. For high schools, this Carnegie unit concept established a standard which, combined with grades in each course, could measure what students were supposed to have learned. Carnegie's reforms ultimately led, indirectly, to the creation of the College Board, Educational Testing Service, the Scholastic Aptitude Test, and the whole accreditation process.

This broader period, from the 1890s up to the end of World War II, also saw the professionalization of almost all the disciplines that we are familiar with in higher education today. As departments of full-time professors formed in the expanding universities and began to grant PhDs along the German Humboldtian model, which combined research and teaching, professors began to form associations to express and protect their disciplinary boundaries. Gary Rhoades, writing in 2009, notes that the Carnegie Foundation was instrumental not only in the creation of standard measures of course work but also the establishment of a nationally portable (but private) retirement system for academics: TIAA-CREF (Teachers Insurance and Annuity Association of America-College Retirement Equities Fund). Through this aspect of the standardization initiative, it helped create the conditions for a "mobile, professoriate that is national in its orientation."[1] This standardization also allowed for ranking of college and universities on a national (and now international) level.

All this took place in one lifetime. This convergence created the "modern" university. The speed with which it happened is suggested by the way the word "modern" came into common use. In the late nineteenth century, in art, architecture, ideology and social behavior including family life, people began to speak consciously of the concerns of modernism. This trend basically continued through World War I and the Depression until after the end of World War II.

EXPANSION

After World War II, the United States held the pre-eminent position in the world, economically, militarily, and hence politically. As military personnel came home from the war, the government passed the Servicemen's Readjustment Act of 1944, or GI Bill (GI actually stands for Government Issue). It originally provided that the government would pay tuition to any institution to which a veteran could gain admission, whether it was Harvard or the local state or community college. The funds not only covered tuition, they included a stipend. The GI Bill functioned essentially as a voucher system for higher education, but without a tuition cap. It was both an attempt to manage a wave of unemployment and to take advantage of militarily skilled, highly motivated but often formally uneducated veterans. One result was a huge expansion in the student population. This population was mostly male. Importantly, the voucher was given to the student, not to the institution.

There was a catch to the GI Bill. Most of higher education was still rigidly segregated. Just as the US Armed Forces fought World War II on a race and gender segregated basis, this sudden expansion of access to higher education did not occur equally across all groups. "Where a veteran could gain admission" left it up to the institution to grant or deny admission. This meant that Black and Jewish veterans were either not welcome or enrolled under quotas in most places, especially in the private schools. A great many institutions restricted women to particular departments and programs, either explicitly or through intimidation. Therefore, though the system expanded tremendously and was somewhat democratized by these changes, opportunities were spread very unevenly.

Many new institutions were established at this time. The 1947 Truman Report, *Higher Education for American Democracy*, was intended to provide broad "education for citizenship." It proposed creating a community college within 60 miles of every American, equipped to offer post-high school as well as vocational training for anyone who asked for it. These colleges began as extensions of K–12 schools, or K–14. At one point in the 1960s, in the United States a new community college was opened every week, totaling over a thousand in the 1970s. As we shall see later, the faculty at these working-class institutions were to become a highly unionized workforce, the highest percentage-wise of any in higher education.

In California, the fastest growing and soon to be the most populous of the states, a broad plan for statewide higher education was developed in the 1950s and passed in 1960. This was the outcome of two contradictory forces. One was the need to regularize and unify a rapidly growing education segment that was diverse, loosely governed, and not really consolidated into a system. It was made up of institutions ranging from the University of California and its campuses to a number of public four-year colleges to various community and junior colleges. The second force pushing toward a comprehensive plan was the need to "cool down" the demand for access to the University of California and many of the state colleges,[2] which were threatened with being overwhelmed by the increasing number of qualified applicants. The answer was what would become known as the "Master Plan," a coordinated three-tier system of tuition-free higher education: the University of California system, the state colleges and universities, and the community colleges. This system would allow access to all three tiers for all Californians and in theory allow motivated and competent students to be assured of the opportunity to progress all the way from their local community college to graduate study at the University of California.

Both nationally and in this system, the demand for faculty was so great that during the entire post-World War II generation, basically anyone with an advanced degree could get a job.[3] People got jobs often before they got their PhDs, sometimes solely on the basis of a telephone call. A great many with only MAs also got jobs, not only in the community college system but also in universities. John Hess, for example, did not have his PhD when he first came to teach at San Francisco State.

Well up to the 1970s, the people hired for these jobs were overwhelmingly drawn from the pool of upper middle- and upper-class white men. These professors would have a degree of prestige and job security unmatched anywhere in the world. For them, the possibility of tenure had become the standard for employment. The origins of this standard can be found in the AAUP *1940 Statement of Principles of Academic Freedom and Tenure* and its predecessors in other organizational statements going back to pre-World War I.[4] The AAUP Statement, while never a source of federal legislation, made its way through the adoption of rules, regulations and policies into almost all institutions of higher learning, as well as into state laws in the public sector. In addition, there followed a long string of case law that established essentially the same protections in the private non-profit section of higher education.

Tenured faculty occupied the contradictory class position of being employees, on the one hand, while at the same time having what was regarded as a proprietary interest (tenure) in their well-paying job.[5] These jobs could easily last for forty years from the date of hire to retirement, and they came with a pension. The class consciousness of this same group, as long as they remained the archetype or norm of faculty, would form an obstacle for union organizing. As the years passed and both the jobs and the demographics of this group changed, the norm would break down but the consciousness, even under stress, would prove to be problematically resilient.

Some working-class ex-GIs, Black as well as white, managed to get through graduate school, bringing their experience with them, and gaining professorships in the late 1950s and 1960s. Although they were a minority, they were to have a large influence on academic politics in the future. Perhaps the most striking example would be Howard Zinn, who became a major leader in the faculty union strike at Boston University in the 1970s, and in the 1980s published the famous *People's History of the United States*. Nevertheless, they did not in themselves make a dent in the professoriate as a domain of middle- and upper-class white men.

There was yet one more important limit on the democratizing impact of the expansion of higher education under the GI Bill: the content of what was being taught, by and large, was not changed to reflect either the new students or the new realities of the world, except in one important way. The demands of the Cold War caused the politically permissible edges of most disciplines to shrink and shift to the right, sometimes quite strikingly. There are numerous stories of professors, both with tenure or not yet tenured, losing their jobs when past association with left groups including the Communist Party came to light. In addition, as science came to be funded by the government under Cold War defense imperatives, the permissible politics of scientists shifted radically to the right from the popular front progressivism of the wartime years.

It was left to the next great transition, the movements of the 1960s, to address what the expansion and partial democratization of higher education by the GI Bill had left undone. By this point, sharp criticisms were being made about what kind of public goods were being produced by this system. What interests were being served? The actions known as "campus disturbances" of the 1960s and 1970s were part of this critique. This included the San Francisco State strike of 1968.

THE MOVEMENT

"The Movement" is the overall umbrella term we are using for the civil rights and Black liberation movements, the anti-war and student movements, the women's movement, disabled and LGBT movements and the various labor struggles such as the farm workers, public employees and the upsurges among manufacturing and healthcare workers.

For higher education, the start of the Movement was in 1960 when four Black students sat in at the Woolworth's lunch counter in Greensboro, North Carolina. They were freshmen at North Carolina Agricultural and Technical College (now part of the University of North Carolina). They were soon joined by students from colleges throughout the South. This sparked a wave of sit-ins in 55 cities in 13 states within two months. Then, stimulated by civil rights leader Ella Baker, these students formed the Student Non-violent Coordinating Committee (SNCC), which would become one of the most important of the Civil Rights groups. SNCC was the core of Freedom Summer[6] and the Mississippi Freedom Democratic Party challenge at the Democratic Party Convention in 1964. Among the demands of the Movement that emerged over the coming years were changes within higher education itself, in terms of curriculum and in terms of eliminating the personal and political restrictions placed on students, especially female students, as well as general access to the institutions.

At this time the wave of children of the post-World War II Baby Boom— more numerous than any generation in US history—began to hit higher education. This generation grew up amid the civil rights, anti-war and women's movements, though not always on the same side. In some cases, they were able to break down admissions requirements in previously selective and restrictive institutions. Yale went co-ed. The City University of New York instituted open admissions, while remaining tuition free.

These struggles went beyond the issue of mere access. They were struggles to transform the curriculum, which had been structured based on the experience of the overwhelmingly white, male professoriate of the time. These professors had created and sustained over generations a curriculum expressing the limitations of their experience and bound by the severe restrictions of imagination enforced in Cold War America since World War II. Now, in the late 1960s and early 1970s, a "new social history" approach developed in the discipline of history along with a revision of the canon in the humanities, the arts and all the social sciences. Ethnic studies programs and departments were created, including the new School of Ethnic Studies at San Francisco State. Experimental colleges, often student-led, and "colleges without walls" were created. Research serving the military and corporations primarily was challenged. Demands were made for the abolition of ROTC

(Reserve Officer Training Corps) on campuses. ROTC programs were the first step in training the majority of officers who served in Vietnam.

In the broader society, these demands took the form of campaigns for voting rights, labor rights, and other social and economic rights and their realization in practice. These rights were now being claimed by groups that had been left out of the New Deal social safety net of the 1930s, such as the Social Security Act, the NLRA and the Fair Labor Standards Act. Women and minorities were the dominant groups among agricultural and domestic workers. Public employees (and among these, of course, were teachers at all levels of public education) were similarly excluded because Congressmen viewed public employees as "their" personal workers. They were also disproportionately women and people of color. In many ways, these struggles in the broader society could be seen as attempts by these groups, twenty and thirty years later, to claim rights that had been won by some white male private-sector workers in the labor wars of the 1930s and 1940s. In the case of Black people, overwhelmingly working class by now, they could be seen as an attempt to re-achieve some degree of functional citizenship (meaning with both political and economic rights) after the defeat of Reconstruction and rise of Jim Crow nearly a hundred years earlier.

Overall, these struggles were successful in the narrow sense of gaining specific reforms especially by expansion and access. But they were not successful in the transformation of the entire society. To the extent that the affirmative action orders of Kennedy (1961), Johnson (1965), and the Civil Rights Act of 1964 were implemented in practice, it was the result of action, watchdogging and enforcement on the ground. In fact, by the 2000s, some aspects of this victory such as diversity among students, especially for Black males, had declined. With regards to the faculty, despite the winning of affirmative action policies on paper at both the state and federal levels, inclusion of most historically excluded groups proceeded very slowly. To the extent that the faculty diversified, it was the growing contingent labor pool that diversified.

As early as 1977, Roberta Alquist Cane, a faculty leader at San Jose State, pointed this out in a memo to the State Chancellor's Office Part-Time Temporary Task Force. The term "Lecturer," which meant a part-time temporary position, had not yet been standardized in the CSUs. Alquist Cane linked the issues of cheap labor and the demographics of the cheap labor pool:

The State College System has been circumventing the entire procedure of due process [to] which all teachers throughout public education in California are entitled. Temporary faculty have been systematically excluded from due process. This has been done by design, for the sole purpose of creating an easily discarded pool of cheap labor ... This is a flagrant violation of the spirit if not the letter of the law which protects teachers from arbitrary dismissal. It should not be necessary to mention that the majority of those victimized by being in temporary positions are women and minorities.[7] (April 28, 1977)

Certainly American capitalism proved itself, along with its higher education system, to be much more flexible than most in the Movement expected. Although some of the activists in the Movement had achieved a revolutionary anti-capitalist consciousness, probably most of them never had a clear idea of how much deeper radical change needed to go in order to achieve the ends they sought.

However, the Movement was taking place at the same time as the political and economic crises of the 1970s approached. On the one hand, there was the crisis of political legitimacy: Watergate, the resignation of President Nixon, defeat and withdrawal of US troops from Vietnam, and a broad loss of faith in government generally. On the other hand, the business sector was seeing a falling rate of profit, general "stagflation" and the rise of economic competitors as the rest of the world rebuilt after the destruction of World War II. The net impact was the destabilization of the system itself.

These combined crises motivated the American elite, together with the elites of its closest developed-country allies, to create a new strategy going forward. This new strategy would replace the "guns" and the "butter," both the war and the social safety net. The name that came to be most commonly used to describe this new strategy was *neo-liberalism*. Higher education, which had until now been somewhat protected from market forces, was going to learn what it meant to be "run like a business."

THE NEO-LIBERAL CONTRACTION

Whether under the labels neo-liberalism, corporatization, or marketization, the roots of this strategy go back politically at least as far as the resistance to school desegregation in the 1950s. In practice, this was racist libertarianism expressed as an issue of states' rights. The background for this is found in Nancy McLean's book, *Democracy in Chains*. Economically, neo-liberalism built on those sections of the business class, especially in the South and Southwest, that had never accepted the accommodation with labor and reform movements of the New Deal, the consequent limitations on their freedom of movement of property and profits, and still longed for the days when property included human beings and the profits that could be made with enslaved labor.

Where neo-liberalism diverges from the classical liberalism (laissez-faire economics) of the Scottish economist Adam Smith, is this: the government envisioned by Adam Smith in the late 1700s was in its infancy as a creator of a public sector. For example, at that time there was still a question of what part of a society was responsible for creating paved roads. Today's neo-liberalism addresses a government that has evolved to play a significant role in social equalization and moderation of the impacts of the capitalist business cycle. Neo-liberalism is "neo-" in the sense that it is a return on new terrain back to liberalism, after the excursion into Keynesianism and social democracy of the first three-quarters of the twentieth century. Under neo-liberalism, the government should only enforce property rights and private contracts, and protect capital accumulation and profits both domestically and abroad. Therefore, the practical functions of government are law enforcement, corrections, the courts (along with the military and some international diplomacy) and the husbanding of the infrastructure necessary to those projects. Other functions are to be reduced and eliminated, or privatized. Even the core functions are to be privatized if possible.

For higher education, neo-liberalism meant to "run it like a business." It brought demands for program-based budgeting and therefore program-based discipline and performance metrics within universities and colleges. Every economic unit within an institution was expected to become a profit center. Small units (like a foreign language institute or a labor education program) might suddenly find that they were being charged rent for the use of classrooms or computer labs. Students became "customers" or even "products" to be tailor-made for corporate employment as demanded by employers, rather than young citizens to be educated. Tuition in the whole sector was pushed up, which represented cost-shifting to students and parents as part of higher education's marketization and this conversion from a social good, to be supported by taxation, to a private good, to be purchased as a commodity, by individual families.

We would argue, along with Richard Moser, that this neo-liberal agenda fully came to higher education only after its main internal opponents, the faculty, had been strategically weakened by casualization. A faculty with tenure, job rights, living wages and benefits, academic freedom, forums where they could influence policy and possess a presence in the public discourse as experts who did not owe any debt to a corporate sponsor—this was a workforce that could still, if organized, wage an effective resistance.

But by the time the neo-liberal agenda came to higher education, the faculty was no longer the same. Despite attrition and retirements, the total faculty headcount had exploded. The percentage of tenured and tenure-track faculty had shrunk, however. New hires were increasingly offered part-time temporary jobs, sometimes called "adjunct positions." In the CSUs, these jobs were called "Lecturers." By the early 1990s, what had been a 70–80 percent tenure-track profession had become a majority contingent faculty, without job security, tenure, living wages and benefits, especially when hired as part-time workers. This workforce did not have the resources, the time, the incentives nor the academic freedom to take risks and oppose the new neo-liberal agenda.

The casualization of the faculty and the great weakening of their collective ability to resist, by loss of job security and academic freedom, was what allowed the rest of the agenda to move forward.

Interlocking challenges lead to casualization

There is no smoking gun like the Powell Memo, that exposed a deliberate decision to implement widespread casualization of the faculty.[8] Unlike some other workforces (for example, taxi drivers) that were casualized on the basis of top-down decisions by government or corporate management, the higher education workforce, which is very decentralized, was casualized through decisions by many middle managers and administrators facing specific problems on the ground. These had to make concrete decisions about classes, course offerings and staffing, while having only a limited view—sometimes very limited—of the overall process in which they were immersed. As middle managers and direct supervisors, they were variously responding to incentives that came down to them from higher manage-

COMPOSITION OF CSUC FACULTY
BY TYPE OF APPOINTMENT, FALL 1972 TO FALL 1976
(Percent of positions filled)

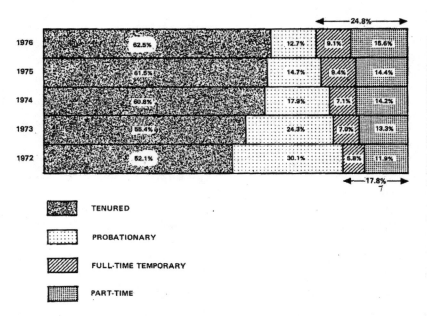

Figure 5.1 Percentages Of CSUC faculty positions occupied, etc., showing rapid increase of part-time/Lecturer positions. An early attempt, 1977, pre-computer and hand-drawn, to promote a common-sense approach to the emerging problem of precarious faculty. It also reveals the persistent, if erroneous, belief that if the full facts of the situation were presented clearly and convincingly to those in power, they would act to change the situation. From the December 1977 *Report of the Task Force on Temporary Faculty* from the Office of the Chancellor, California State University and College.

ment, trustees, legislative bodies, corporate funders and other donors and major actors in their community. These incentives created problems which managers at the local level had to solve in the context of their actual realities.

The actual realities that they faced at this time were (and are) an increase in demand for higher education, an increase in credentialism and therefore a demand for degrees by employers, a much greater curriculum breadth both inherited from the struggles of the 1960s and the demands of employers, and the need for more and different student services than had been required in the past.

One way to consider this complicated history is to divide into four types the managers' problems for which the casualization of faculty became their main solution.

"Do more with less (money)"

One problem for managers was that government appropriations for higher education began to decline throughout the 1970s. This was both for political reasons, such as pushback by state government against the disruptions of the student movement, and for economic reasons, such as stagflation and the continuing shift of the tax burden away from corporations and the wealthy to the working people and the middle class. The net result was less money for higher education on a per-student basis at the same time as demand remained high. In the private sector, the profit crisis, which was key to causing the neo-liberal strategic change in the first place, also caused charitable donations and the investment revenue from endowment funds, which had subsidized private higher education and kept its tuition manageable, to shrink or at least stagnate in purchasing power.

In both cases, public and private, there was a strong incentive for anyone in higher education who had a budget to manage, from department chairs to chancellors, to lower labor costs. The number of students required to generate a full-time budgeted faculty position in the CSU system went from 16 to nearly 18 in 1976–77.[9] They could do this by expanding the once-small group of part-time contingent faculty and separating their pay rate into a permanent second tier. These faculty were now significantly cheaper to hire. Not only were they often paid as little as one-third of what tenure-track faculty could command, but they also often did not receive employer-paid benefits or retirement contributions, costs that were rising rapidly and could make up a third of a total paycheck for a tenure-line professor. So hiring contingents started as a temporary response to an economic crisis, but it rapidly became the new normal.

Unpredictable enrollment patterns

A second big change was a radical shift in not just the size but the composition of the student body. In the years post-1975, all of non-elite higher education saw an increasing percentage of students coming in the door who were working adults, many with years of experience in the labor market. Non-elite higher education is counter-cyclical with regard to unemployment and the business cycle. People go back to school when it is hard to find a decent job or they are worried about being laid off or their skills being made obsolete. So people were returning to school to re-train or add new skills or start new life courses.

Many of these students could not afford, either economically or in terms of their adult life commitments, to attend school full-time or consistently, despite their serious motivation. Their attendance was contingent upon jobs, financial aid, low or zero tuition, the availability of childcare, healthcare, accommodation for physical or psychological disabilities (which especially affected veterans) and the availability of classes when and where they needed them. This new situation was a tremendous headache for lower-level managers (assistant deans, program chairs and department heads) because it meant they could no longer predict how many sections, what courses and at what times, locations and have them fill but not over-fill. "Filling" of course is an arbitrary measure, and class sizes have increased ever since this period.

In California, the cutbacks caused by Proposition 13 (1978) and other anti-tax campaigns slowly drained the K–12 system of resources. Class size increased. Remediation was passed upward from high schools into community colleges and the four-year institutions as an obvious consequence of this starvation. Colleges had to set up and expand remedial reading and writing programs. Computers entered the classroom and learning labs, using fewer, cheaper staff, burgeoned.

Just as the first problem pointed to the economic crisis, the second problem pointed to a flexibility crisis. Managers could no longer look at the regional high school graduating classes and predict how many sections of English 101 they would need to offer in the coming semester. The answer, first for the short term and then in the longer term, was a flexible, contingent faculty that could be hired on a per-class basis and laid off ("de-scheduled") when not needed. They would be in no position to complain about the gradually increasing class size. Somehow this seemed ironically appropriate: a just-in-time faculty for just-in-time students.

Unionization as a threat to management control

An important aspect of the impact of the social movements of the 1960s and 1970s on higher education, besides the demands for changes in administration and curriculum, was the rise of a union movement among faculty, graduate assistants and other campus workers. This was co-terminus with the general rise of public-sector unionism, founded largely with the inspiration of the civil rights movement and with much the same spirit. This spread into other service occupations like healthcare and welfare work. The passage of HEERA in California in 1978 was part of this rise.

The creation of a permanent two-tier system within the faculty was a powerful weapon against the emerging faculty unionization movement. Not only did it divide the faculty, it forced tenure-line faculty to confront questions such as: who should be in the union? Are these fellow-faculty potential brothers and sisters, or are they scabs against whom we should try to build a wall? (The word "scab" was actually used to describe non-tenure-line faculty in some real-life cases. "Unprofessional" was another.) As we have seen, this very question roiled the faculty workforce in the CSU, leading up to their collective bargaining elections in 1981. At the very least, a two-tier system was a serious distraction from the struggle against the employer. At the worst, it resulted in union drives that were hamstrung, weakened and often aborted. The two-tier system also resulted in actual unions that were sometimes incapable of either representing their emerging contingent majority or exercising the strength and authority vis-à-vis the employer that the situation called for.

This is not to say that most managers saw this divide-and-conquer opportunity clearly at the time. Nor did they have it as a major conscious motive for casualizing their own faculty. Certainly, some did. But once the benefits of this tactic began to be apparent, it was then clear to all that it was an effective strategy to resist unionization. It is one that is now almost universally supported by management, as shown by their consistent legal challenges to attempts to create new combined tenure-line/contingent bargaining units.

The newly diverse hiring pool

Problem four that casualization solved for managers was more political and only indirectly economic. Of all four arguments, this one is the trickiest to make and the most provocative. Basically, the argument is that the lower-paid, lower-status positions made available through casualization could be justified because now a larger percentage of the people hired into those positions came from the newly diversified faculty labor pool that contained relatively fewer white men and more women and more minorities—that is, more lower-status people.

Under the pressure of the various movements demanding access to higher education, the undergraduate student body was becoming substantially more mixed and diversified racially and in terms of class, age, gender and disability status. Gradually, this change made its way into the graduate schools. This was where the future teachers of higher education were trained. By the 1970s, a mixed cohort of potential faculty was in the job market. These people were fully qualified, in many cases *more* qualified than their predecessors. Because they were young, these new faculty were much easier for students to relate to as role models. They were likely to be popular on campuses. They might even develop "a following," which was threatening to senior faculty and managers. Also, many of them were veterans of the movements of the 1960s, so they brought a distinctly different outlook to their work. Crucially, though, many of them were not white men of upper- or middle-class background.

This allowed an argument to develop that essentially said, "Since a lower class of people want these jobs, the jobs themselves are clearly lower class than before." The fact that an increasing percentage of the students taught by these lower-class professors were also non-elite, working-class people of color and/or women added power to this argument. Politically, it became easier to argue that these jobs did not have to come with the traditional perquisites—compensation, authority and respect—that the previous generation of white male college professors had enjoyed. Women, after all, were just working for "pin money," and minorities were presumably "used to a lower standard of living" anyway.

So newly hired contingent faculty, armed with their MAs and PhDs, found themselves teaching classes of ill-prepared open admissions student in underfunded institutions. They were hired into positions that lacked office space for private consultations, payment for office hours, professional development support, the power to consult and advise on course and program design, and many other deficits that readers of this book are familiar with.

In other words, they were without the very resources needed to perform the services that the expanded student population desperately needed.

Managers thus found themselves in an awkward position. This was, on the one hand, the period of affirmative action. The need to hire faculty who looked like the more diverse student body was clear. But on the other hand, the very diversity of the labor pool from which the new hires might come made it easier to degrade and increase in number those second-tier jobs. By doing so, managers were able to use the victories of the Movement to maintain and even increase their control over the faculty and ultimately the institution as a whole, weakening the tradition of faculty governance.

This transition of a job's status to match the stereotype of the person doing that job is not unique in the history of work. In fact, it was a repeat of the historical change in office work. Being an office clerk or a secretary started out in the mid- to late 1800s as a job that aspiring executives took on their way up to management positions. As women entered that expanded work-force through their secretarial training and the invention of the typewriter, it became a woman's glass-ceiling job: low wage, low respect and without benefits. Labor costs went down and capital was invested in the typing machines themselves, not the typists, who were actually called "typewriters."

These changes also meant new challenges for faculty unions for building solidarity on a level that had not been confronted in the past, when the faculty was overwhelmingly white men of middle- and upper-class background who shared a common culture and sets of assumptions and privileges.

THE NEW REALITY

These four problems drove managers toward solutions that regularized the use of contingent faculty in degraded positions and the creation of a seemingly permanent two-tier system.[10] Just at the moment that the faculty in higher education were creating their union movement, this new reality was solidifying. The solutions interlocked to form a seemingly single and irreversible reality.

To deal with this new neo-liberal reality, contingents and higher education faculty unions were going to require a different analysis and strategy. In response to this need, in the late 1990s and early 2000s, a network began to form out of what had been a scattering of contingent organizations across the country. This network would become the basis of our national and then international movement.

Looking back across these four transitions, it is easy to see that they have come with increasing frequency. Roughly speaking, the transition to standardization took two generations, the transition of expansion was 15–25 years, and the period of the Movement was 15 years, but overlapped with the neo-liberal contraction which has been the dominant trend in the last 40 years. Each transition represents the outcome of conflicting social forces in an increasingly important terrain for society. The centrality of this terrain to the concerns of society may explain why these transitions are coming about more often, but in every case, they reflect changes taking place in the broader society that redefine what Clark Kerr identified in the title of his book as *The Uses of the University*.

In the next section, Part III, we will examine what we as contingent faculty might fight for in a fifth transition.

PART III

WHAT WE WANT AND WHAT THE CFA GOT

6

Blue Sky #1 Organizing and Economics

At the end of Chapter 5, we described four problems faced by local-level managers and administrators that could each be solved by the single solution of casualization: lack of funding, unpredictable enrollment patterns, unionization, and the increasingly diverse labor pool from which faculty could be drawn, this last being both a problem and an opportunity. We pointed out that reliance over the decades since 1970 on casualization as the one-shot solution locked in a seemingly intractable, irreversible management strategy. Nor could casualization be challenged by offering alternate solutions to each of these management problems separately. More funding would not necessarily go toward hiring faculty. Arguments based on pedagogy for supporting classes that might have momentary low enrollment or for supporting "marginal" programs that answered the needs of minority populations would fall on deaf ears in an economic demand-driven, much less a profit-driven system, compared to social need-driven system. There is no effective alternative to unionization, however, and the answer to the diversity "problem" has to be equality—easier said than done.

Therefore we have to deal with casualization as a totality, not by offering a set of separate solutions. We need to fight back against it on a different basis: on the basis of our own goals as teachers in higher education. We have to ask what higher education should do and what we need in order for it to serve that purpose.

Our overarching goal is to abolish contingency and precarious work as a condition of our lives and the lives of all workers. Socially useful work, including our work, should carry with it security of employment commensurate to the social need for our work. This includes, for all workers, freedom of association and speech, a living wage, appropriate benefits, and the opportunity to choose to do this work on a full- or part-time basis for all who are qualified and ready to do it as long as the need for the work (not necessarily the economic demand) exists.

We now put forward a blue-sky list of the goals that this movement fights for, along with some of the differences that have arisen over these goals. In addition, we will note the various terrains where these fights take place, often in more than one setting at a time.

GOAL 1: THE RIGHT TO ORGANIZE
AND THE RIGHT TO BARGAIN

The first goal we fight for is our right to organize and to be recognized. We have to fight for the political space and the legal right to organize a union that can represent us.

The approach we take to organizing prepares us for the struggle down the road. This means that we have to organize for the right to organize. We have to act like a union before the government or even the employer recognizes us as such.[1] In this, we confront questions about what kind of union we want. Do we organize a service-model business union, or a class-struggle, rank-and-file, or social movement union? Do workers organize their fellow workers, or do staff organizers do that job? Do we organize by bringing people together and acting like a union, or by focusing narrowly on the legal right to organize and winning in court? Do we organize by promising immediate economic benefits or by holding up the hope of being part of a movement? The future of our union can be forecast in the way we organize at its beginning.

Although examples exist where a group of workers in higher education have self-organized and forced their employer to negotiate with them through direct action, without the protection of enabling legislation, they are not common. Examples of this include the faculty union in the 1968 San Francisco State strike, the Chicago City Colleges in 1965 and graduate students at the University of California Berkeley (1964) and then at the University of Illinois in the early 1990s. Most recently, in 2018–19, we saw a wave of teacher strikes in "red" states like West Virginia, Oklahoma, and Arizona that lack collective bargaining laws for teachers, which were able to force negotiations, in this case at the level of state government. In some cases, teachers have done this not only without the backing of the law, but explicitly against a law that forbade their striking.

The predominant way to bring an employer to the table and negotiate is by organizing a union and getting it certified by the pertinent labor board. For the private sector, this would mean the federal NLRB, or in the public sector, a state labor board, if one exists under enabling legislation. Either way, the fight for the right to organize goes on in at least two arenas besides the legislative. The first arena is against the employer, since the employer will almost always resist an organizing attempt. The second is less obvious: it is against the ideas that have been spread among workers who usually know very little about unions. They need to be educated and then relieved of their fear, to the extent possible, by being brought into some collective action. The fight for the right to organize, in other words, takes place across the whole of the social world that is connected to the workplace, from dinner tables to coffee shops, from boardrooms to legislatures, and from classrooms to offices.

People virtually never fight for the right to organize in the abstract. They fight for it when it is the only way they can win on issues that concern them at the moment. So we define the terrain of organizing and we win the right to do it by doing it, acting in solidarity for the demands that large numbers of workers feel are most pressing at any given moment.

THE TERRAIN OF COLLECTIVE BARGAINING

The right to organize comes before—though not by much—the right to bargain. But first we must take a step back and talk about the terrain of collective bargaining.

Collective bargaining is where we actually engage with the employer. It can lead to contracts, which are agreements that can be enforced by grievance procedures, law, and/or our own direct action. When we fight for the right to collectively bargain, we should fight for as broad a scope of bargaining as possible. This is a fight all by itself. Employers, once they have been brought to the table, will want to narrow the scope of bargaining down to a minimum, to wages and working conditions defined as narrowly as possible. We, on the other hand, should organize to make sure that the scope of bargaining is as broad as we can make it, including the full range of conditions that make it possible to do our jobs well. This means bargaining for levels of control over our work.

Another way to talk about the scope of bargaining is to divide issues into two categories: economic and non-economic. We understand that all economic demands have non-economic implications and vice versa. But these categories are how people on both sides of the bargaining table generally think of them. Things that fall into the category of "economic" are wages, benefits, and anything for which the cost is easy to calculate: office space, computer equipment, release time, conference support, and so on. Things that fall into the category of "non-economic" usually raise the issue of control. Anything that takes discretion away from the manager—in hiring, seniority rights, credit for previous service, scheduling, assigning curriculum, even something as simple as access to photocopying—and gives it to the employee is about control. This is why these "non-economic" issues, especially job security, are so fiercely contested. Often more difficult to win than economic issues, they are also issues concerning dignity and "respect."

Other unionized professions often make clear explicit connections between control of their work and social needs. Nurses, for example, have invested heavily in political campaigns that favor universal healthcare, Medicare for All, and workload and overtime limits. Less publicly, the building trades (sometimes) have led in practicing and teaching environmentally sustainable construction. Firefighters push routinely for adequate fire safety and fire prevention laws. But the relationship between the loss of academic freedom and the rise of "anti-science" or "climate change denial" and "fake information" is not always clear. For faculty, control of their work is closely related to academic freedom, which is in turn related to the social justification for the higher education system at all. This is why broadening the scope of bargaining is a fight, and control of our work is something that has to be fought for. In higher education, it is usually easier to get employers to agree to economic issues than control issues.

It is also important to make clear that collective bargaining is a relationship of orchestrated power struggle. This is just as true for faculty as it is for other kinds of workers. Bargaining is not an Oxford Union debate where the most logical argument is awarded a prize by a third-party judge. It is not a collegial conversation between equals. The most agreeable negotiator does not get the best contract. It is a power struggle.[2] If war is politics by other means, to paraphrase von Clausewitz, direct action is a continuation of collective bargaining by other means.

What we win in collective bargaining, both in the scope and the content of our agreements, is a temporary truce represented by, in American parlance, a "union contract." Both the scope of bargaining and the content of the agreement are merely a reflection of the relative power balance between us and the employer at that moment. In the case of public employment, it is a reflection of the struggle between us and not just our immediate employer, but of capital as a whole, mediated through the state.

GOAL 2: A DECENT WAGE

The single most important economic factor that we bargain for is clearly the wage. Whether it's called the hourly wage, a salary, an honorarium, a per-class stipend, or any of the other euphemisms, it is the pay we receive for our work. Contingent faculty pay in the United States can be as little as 20 percent of what a tenure-line faculty member would make for comparable work. Wages are even worse among online faculty. There were still per-class adjuncts in the United States in 2019 working for less than $1,000 per three-unit class (and at the richest, most elite universities, sometimes even for free, simply for the prestige of the association with the institution). At the other extreme are cases where the contingent wage is within hailing distance of a tenure-line wage on a per class basis.

The best compensation system, for us, is usually pay by percentage load, linked via a high pro-rata factor to a single salary schedule for all faculty, including tenure-line faculty. For example, take a college where a full-time load is four three-unit classes per semester. At that college, a contingent teaching two classes per semester would be counted as having a 50 percent load, so if all faculty were on a single salary schedule, a contingent with the same experience and qualifications would earn 50 percent as much as the comparable full-timer. This would be called 100 percent pro-rata pay. If the contingent had no committee duty or research requirement, and there was a 20 percent salary discount because of that, the contingent would have what would be called 80 percent pro-rata pay. To further equalization, lump sum raises are preferable over percentage raises.

In Chapter 2, we wrote about Elizabeth Hoffman and the long effort to defeat the various forms of "merit pay" which were typically proposed by management with the incentive that they would reward individuals for excellence, special service, etc. Even when "merit pay" is not proposed as a substitute for some kind of standard, secure, salary increase that applies to everyone, it can still be used to express favoritism and discrimination and to divide the faculty. Therefore any kind of discretionary salary increase should be discouraged.

GOAL 3: BENEFITS, BOTH AT WORK BUT ALSO FOR EVERYONE

After wages, the second most important economic factor we bargain for is benefits. These benefits used to be called "fringe benefits," but they are of such substantial importance now that "fringe" has dropped out of common usage. Without a national health service program, employer-paid health insurance and benefits are, at many moments in our colleagues' lives, the most important aspect of the job. When a teacher or member of the family gets sick or injured, the benefits package can mean the difference between life and death. Some contingent jobs only offer benefits during a semester, with nothing provided between semesters or over the summer. We can bargain over conditions like this, too.

Although we fight for healthcare benefits at the bargaining table, we understand that the vast majority of contingent faculty will never achieve adequate healthcare reliably through employment. Therefore, along with the nurses and others, we fight for universal healthcare decoupled from employment. This is an example of an issue that we fight for on multiple terrains at the same time.

Another benefit of great importance that many of us do not have and have to fight for is a retirement pension. One contingent famously said, at a meeting with other contingent faculty, "My retirement plan is a bullet in my desk drawer." Others—including some well-known contingent movement leaders—continue teaching well into their eighties, unable to retire. In some states, we are covered by a public state retirement system. Access to these plans is often on a discriminatory, second-tier basis, where our work is not fully credited and our time-to-vesting is unfairly lengthened. The benefit formulas often discriminate against us because of our low and inconsistent pay and multiple employers. In other states, and in the private sector, we may not be covered by a pension plan at all except for Social Security, if even that.

In 1988–89, the Consolidated Budget Reconciliation Act (COBRA) included a requirement that all part-time higher education faculty be covered either by Social Security or by an "equivalent-adjudged system." Many contingents working at that time had not been contributing to Social Security or other retirement systems at their teaching jobs. The "equivalent-judged" system exception gave some states and even individual employers the power to opt out of Social Security and design a system more subject to local control. "Local control" here meant exposure to the ups and downs of state legislatures and budgets, usually to the detriment of contingent faculty. A thousand flowers—or weeds—bloomed in that space.

An example of one of these "weeds" is the elimination of defined *benefit* pensions like Social Security and most public teacher retirement pensions. These are plans where the amount of money that a pensioner receives is set, known, and usually indexed for inflation ahead of time. These plans have been in many cases replaced with a defined *contribution* plan, where the worker contributes into the plan but then is completely subject to the vagaries of the stock market, in which the money is invested. The employer may or may not match it at all. This passes the risk of the market to the pensioner. These defined *contribution* plans are known as "401(k)s" or for most of us in education, as "403(b)s," also referred to as "TSAs" or tax-sheltered annuities, and named after the section of the US Tax Code where they are found.

This change over recent decades in both the public and private sector has benefited employers, enabling them to lower the amount and cut the percentage that matches what we contribute through payroll deduction. An employer's current contribution under Social Security is 6.2 percent for covered employees, plus 1.45 percent for Medicare. In states where bargaining over non-Social Security pension contributions is allowed, some employers have managed to negotiate their contribution to the plans down to zero.

An example of someone laboring quietly in an unappreciated but profoundly valuable corner of the contingent faculty movement would be the person at a campus who researches the retirement plans practiced by their employer and then teaches and advises other contingents on how it will affect them. These colleagues are among the activist heroes of our movement.

Regarding Social Security itself: some of us are directly covered by our teaching jobs, depending on the state. Some of us are covered by public or private systems that have been adjudged "equivalent." Some are covered by both. If you are covered by Social Security, it will show on your paycheck as FICA, referring to the Federal Insurance Contributions Act, which is your half of the payroll tax for Social Security and is matched by your employer. Even if you are not covered by Social Security on your teaching job, if you have sufficient earnings in forty quarters of some other covered employment, you may be able to claim a Social Security pension when you reach your age of eligibility, and Social Security disability insurance (SSDI) and/

or survivor's benefits. This is separate from Medicare, which is another deduction from your paycheck and is supposed to be universal, whether you are having Social Security deducted or not. These are different deductions, even though both programs are administered by the Social Security Administration.

There are two things left to explain. The first is that 13 states, including California, Texas and Illinois, provide public pension coverage but not Social Security for teachers in general. Details regarding contingent faculty vary by state. The second thing is that, in a misguided attempt to keep a relatively few top-paid executives and administrators of agencies from double-dipping unconscionably high pensions when they combined their Social Security with their state public pension, Congress passed a number of laws to limit these "windfall" pension combinations by a formula deduction from their Social Security pension.[3] The unfortunate result has been that teachers who qualified for Social Security have found their Social Security artificially reduced by this formula. For us contingents in the 13 states, it has meant a substantial reduction in what for most of us was already a small Social Security pension. All of the teacher retirement organizations and education unions have been lobbying against this for years but so far, to no avail.

The purpose of going into the issue of retirement benefits is to emphasize that, like medical benefits, they are something that you don't need until you need it, and then when you need it, it can be a matter of life and death. Most importantly, a 401(k) is not a defined benefit plan; it is little more than a savings plan invested in the stock market. We should understand how these plans work, and we should be prepared to fight effectively for the right ones at the bargaining table, in legislatures and public agencies.

Since the economic situation of contingents is so different from the economic situation of tenure-line faculty, we need to have well-prepared spokespersons negotiating at the table, who are actually contingents themselves and can give the arguments the importance they deserve based on having lived the experience.

A number of other benefits to which we can pin on a dollar sign, and that we fight for, are generally taken for granted by for our tenure-line colleagues. These include paid leaves like sabbaticals, money for professional development and travel, tuition remission for ourselves and family members, access to recreational and wellness facilities, and free or subsidized parking and other transportation subsidies. There is also pay for non-classroom work, either as part of equal pay with tenure-line faculty or by the hour spent. This would include pay for office hours, department and other meetings, curriculum development, other class preparation, creating online courses, class cancellation fees, etc. Another economic issue is sick leave or maternity/paternity leave. In a few states like California, contingent faculty have sick leave by state law, at least in the community colleges. But in most states, this has to be fought for either legislatively or at the bargaining table. In any case, how it is implemented—for example, the calendar for the Family Medical Leave Act (Federal) (FMLA) eligibility or its expansion—and anything in addition to state requirements can be the subject of bargaining.

GOAL 4: OTHER ECONOMIC ISSUES

Four other economic issues are important for contingent faculty: access to unemployment insurance, worker's compensation, state disability benefits, and adherence to occupational safety and health standards.[4] These issues span economic and legal questions in ways that are particularly significant for contingent faculty. Therefore, they need to be fought for on various terrains. These are rights that come from court decisions, administrative regulations, or legislation that has been passed into law, but which depend on the cooperation of the employer who has a financial incentive to avoid making them happen. The more complex and drawn-out the process, the more likely the worker who is trying to claim these rights will give up and go away. Too often we hear, "I applied but I was denied." This is the first-line response from unemployment benefit administrators and simply means that the applicant needs to appeal, after getting advice.

We can influence these rights and benefits through bargaining. We can specify and reinforce that a worker has such and such a right; we can also bargain the efficiency of the process and the timelines under which this happens, as well as the right for the union to have a person representing the worker in claiming various benefits and protections.

Good resources are available on the Internet that give explanations of all of these. However, most of these resources assume that the subject is a full-time permanent employee. Interpreting them to apply to our situation requires judgment calls. This is why we should put pressure on our unions and organizations and engage with the various agencies and nonprofits that are associated with these rights. An example would be to ensure that a contingent faculty member trained in understanding unemployment issues is available to advise contingents during the application process.

Now we will turn to issues that are generally considered non-economic factors, also called "professional issues."[5] These include the essential goals of job security and academic freedom.

7

Blue Sky #2 Job Security, Academic Freedom and the Common Good

The goals in this chapter are often described as non-economic factors—although everything can be shown to have an economic side. As we mentioned above, these issues are often about control and therefore also about dignity. They are therefore inevitably harder to successfully bargain over. We begin with two of these goals that are inextricably bound together: job security and academic freedom. Without job security, we do not have academic freedom. Without academic freedom, we cannot fulfill our role in society or our duty to our students.

GOAL 5: JOB SECURITY

By far the most important non-economic factor is job security. For many of us, this is even more important than the compensation issues. The essence of contingency is obviously job insecurity; our insecurity is the employers' flexibility. These and many other non-economic factors are directly related to the power the employer has over us in the workplace. Therefore employers often resist much more on these questions than on strictly economic ones. Money is fungible, but power is absolute (up to a point).

Job insecurity can be divided into two aspects. One is security in the job we already have, whether it is full time or part time. The other is access to tenure-line employment, the promise of increasing security. We will deal with these separately.

Security in one's current job

We want to know that we will maintain the level of employment that we have at a given moment through the current semester and into the next. This means a degree of job security in the job we have now. This immediately raises the issue of seniority and leads to bargaining for a seniority system. This is also the single most contentious issue at many bargaining tables. Establishing a seniority system can be the first "spanner in the works" of contingency itself. For this reason, managers on the other side of the bargaining table resist. Even the very word "seniority" itself is often resisted militantly by the employer. When we look at the CFA/CSU system contract we will see that although this contract does provide significant grievable seniority rights in practice, the word "seniority" actually does not appear anywhere.

There are many different ways of defining seniority and seniority systems, all of which have to factor in what constitutes proper qualification to teach a class or perform some other duty. The question of what constitutes proper qualification is a topic in itself. The power to judge and critique the quality of academic work, including the worth or worthlessness of paper qualifications or credentials, is something that historically belonged to faculty, not to non-faculty managers who were hired to administer the institution. Faculty are the right judges of faculty. This is the very principle behind peer review of all faculty work, teaching, research, data management, etc. It is also the principle behind the existence of faculty committees, faculty shared governance, institutional review boards and ultimately institutional autonomy. This means that the question of what constitutes a proper qualification must be left to the faculty and not ceded to the managers. The institution's autonomy depends on the capacity of the faculty to carry out their role.

Within this inheritance from our feudal past as a guild may lie the seed of what a now-proletarianized faculty can build upon for a twenty-first-century system of worker control of higher education institutions—that is, our workplace.

What constitutes proper qualifications and who decides this question should be bargained. In other words, when the question of seniority comes up and leads to the question of qualifications, the faculty should not only design and propose such qualifications, but also bargain for an agreement on it. One problem is that in too many cases, when designing a system of qualifications to be administered by faculty, the definition of "faculty" at the time does not include contingent faculty. As a result, the qualifications are often slanted to exclude us. The same problem exists regarding evaluations, which are an element within qualifications. We will see how this is handled when we look at the CFA/CSU contract.

Seniority systems can be based upon a variety of factors—for example, the first date-of-hire at the institution, maintenance of a relative load, or seniority of teaching a particular course or subject. Seniority can be by program, department, school, or institution-wide. The way each type of seniority intersects with the issue of qualifications can vary. The assignment system that flows from various seniority systems can vary as well. There is also the question of multi-semester contracts which can be either full- or part-time, and the selection process by which people get them. All of these are bargainable. For readers who are preparing to bargain their first contract, these alternatives will give you a headache but it also may be the most important work you ever do.

The best seniority system for the greatest number of people and for our collective power is still a matter of debate within the movement. No matter how complex these questions become, the point is that they can be debated within the union and contingent faculty bodies. Again, management at all levels tend to resist any interference into their discretion. We should have transparent and democratic discussion when a proposal is brought to the bargaining table. Anyone with a stake in bargaining over seniority should consult a wide variety of existing seniority systems when trying to create one or improve one that exists. Good examples are not limited to faculty contracts; we can learn from the practices in other industries. Useful ideas about seniority systems can be found in manufacturing, like the auto workers (UAW), steelworkers (USW), in entertainment, in longshore (with their seniority hiring hall), and especially in industries where workers are hired by the project, or seasonally.

Increasing security over time: Access to tenure-line employment

Now we turn to access to tenure-line employment, the promise of increasing security. One current term for this is "upgrading." This second aspect of job insecurity for contingent faculty can also be bargained. It is difficult to do so in places where there are separate bargaining units for contingent and tenure-line faculty because management can argue that it is a concern for the tenure-line unit, not for the contingent faculty unit. But in a combined bargaining unit that includes both contingent and tenure-line faculty, the issue of access to upgrading can be and has been bargained.

Where this has happened and actually been implemented and policed, it has resulted in a sea change over time both in the culture and in the respect accorded contingent faculty within the bargaining unit, the union, and the institution. The most prominent example is the City College of San Francisco AFT 2121 contract. There, the technique used is "first consideration." This means that the employer must consider hiring the inside candidate first, not just chronologically but also in the sense of "first choice." Subsequent arbitrations have made explicit that the employer has to demonstrate that any outside candidate is better than—not just equal to—in qualifications for the job in order to be hired. (Exceptions can be made for affirmative needs.) Priority for the most senior contingents can also be bargained.

Then there is the question of access to the rapidly growing number of non-tenure-line full-time jobs. Like the question of qualification, the question of who is chosen from the contingent pool to get a multi-semester contract or a full-time non-tenure-line job or a tenure-line job—or to be interviewed for any of these—is bargainable. In quite a few places, contingent faculty have at least bargained guaranteed interviews.

The issue of trying to bargain for more actual tenure-line jobs—like the issue of combined bargaining units versus contingent-only bargaining units—remains a controversial one within the contingent faculty movement. Many contingents argue that establishing more tenure-line jobs or even more full-time non-tenure-line jobs primarily provides management with the ability to cherry-pick favorites among existing contingents or newer, often younger outside applicants, while at the same time inviting them to lay off older, more experienced, better-paid, and troublesome contingent faculty. There is no shortage of examples of places where this has happened. On the other hand, schools that have the highest percentage of faculty as full-time tenure-line and have combined bargaining units are among those where the other conditions for contingents under their contracts are the best. Many argue that combined bargaining units allow the potential for faculty to work out legitimate criteria for a seniority- and qualification-based system to allocate these jobs while at the same time protecting the rights and jobs of existing contingents. This has proved difficult in many places in practice, but where successful, as it was at City College of San Francisco, it has achieved the best conditions. To the degree that a fair seniority system is implemented for both upgrading and continued assignment, it also lowers the level of competition among faculty and therefore strengthens the union as a collective body. This is yet another reason why managers so strongly oppose the creation of seniority systems.

Other important non-economic goals

Among the other professional or non-economic goals that can be bargained (although often only as non-mandatory subjects), some are collective, like the number or percentage of full-time tenure-line jobs, and some impact everyone as individuals, like access to an office. Access to secure and private offices for doing work and consulting with students and adequate equipment (telephones, computers, printers, Internet accounts) and supplies (paper, red pens, even chalk, etc.) are high on any contingent faculty wish list. The inclusion of faculty names in schedules (rather than listing classes as "TBA" or "staff") and on the Internet, not to mention on doors, mailboxes, and wall signs, can also be bargained. Staff support, in the form of computer and technical support, including for faculty who have to maintain their own technology from their homes, is also a bargainable issue.

Some bargaining issues can be understood as a pushback against the way managers have used faculty "flexibility" to solve their problem of unpredictable enrollment. For contingent faculty, improvements in stability and continuity of employment, even in small increments, are important. Early dates of assignment can be bargained, to discourage the assignment of classes only two or three days before the beginning of a semester. Under the current conditions of austerity, administrators also often try to impose higher class-load minimums and maximums, causing class cancellations, as well as artificially cancelling classes early, thereby causing contingents to lose employment and forcing students to move into another section, endure a larger class size, or often drop out entirely. Some unions have bargained what is known as "bump insurance," or class cancellation insurance, where contingents get paid a bargained amount when they lose a class within an agreed-upon time frame before the class would have started or after the class starts. Collective bargaining can actually rationalize the work of a faculty as a whole by creating deadlines and forcing a routine process that is often just delayed by administrative procrastination and insensitivity.

Access to resources for professional development, time off—not just remuneration, but also credit for doing professional development or providing it for fellow faculty—and time to organize and consult among our colleagues are provided without question for tenure-line faculty. These are key aspects to being a full member of an institution and able to contribute to the culture of the institution and the collective professional development of the faculty and staff. This also means in practice the right to participate in practice in governance structures, with a full vote, and preferably to be paid for it. This has proven to be bargainable in many institutions, and is a way to encourage the building of the collective knowledge among us that enables us to do our work together better, for ourselves and our students.

Elizabeth Hoffman pointed out that anything that suggested permanence, from parking lot passes to mailboxes to direct deposit, was likely to be resisted by management that relied on employee flexibility. The status "emeritus" is, of course, one of these. She tells this story of how it happened at Long Beach State University, where she taught:

We went after emeritus policy and got it. We tried and they said we couldn't so we said, we'll grieve that. We got it but they weren't telling people they could get it, so one of the things we learned was that we had to bargain some notification stuff. So now, if you go to a Dean or a Department Office, if you have a letter that says Elizabeth Hoffman now has X, they have to give it to you—they are cowards.

She tells the story of what motivated her to make this fight:

The reason I finally got an emeritus card was because my daughter went to Long Beach. That was about twelve years ago. She started as a freshman. She had been hanging out at CA Long Beach since she was a kid. She was at Orientation and they handed her the catalog and they told her all the faculty was there and she looked and could not find me. Now she had seen me going to work at Long Beach every day of her life, and hanging out there. So, why aren't you listed? My mother has been lying to me all these years! Her lights were blinking! You've been working there for twenty years, right?

So I looked in the catalog and I wasn't there. Not many of the Lecturers I had been with were there. It's an academic senate policy and it's automatic to get in the catalog—so I asked the academic senate chair and couldn't even get a response. Finally she said, "It's our practice that Lecturers do not get in the catalog." So then I went to the Faculty Affairs

Committee (they deal with tenure issues) and said, "You cannot let them ignore the policy that the Senate put in." Then I went to the union and they wrote a strongly worded letter. Then the Faculty Affairs Committee chair wrote back and said they had that practice but not until [the] next year because they did not know who the Lecturers on the list were.

I said "You have to give them paychecks, don't you?"

Well, I had to go down to the parking lot to get my free emeritus parking pass. There was a student there in the kiosk. I said, "Can I please get my emeritus parking pass?" And she said, "Oh, let me look you up on the emeritus list, oh, here's your name, sure." So if you want to know who is on the list just go down and ask the student worker in the parking lot because she's got the list.

GOAL 6: ACADEMIC FREEDOM

The final non-economic professional question, after job security, but linked to it, is the most important. This is the question of academic freedom. Our lack of academic freedom flows from the insecurity of our jobs. The employer's flexibility can also deny us the effective right of free speech on the job—in the case of academia, academic freedom—that should be the essential part of any faculty member's body of rights and responsibilities.

While academic freedom is especially important to us as teachers and researchers, it should not be seen as a special perk for a privileged guild. It is in fact another form of workers' right to free speech on the job. "Free speech on the job" is linked to the public's need for a worker's right to speak openly about their area of expertise. The recipients of all goods and services also need to trust that the producers of those goods and services are free to speak to truth as they see it. The example closest to teachers is healthcare workers, but it is easy to see that this applies to construction workers, transportation workers, food and restaurant workers—the list goes on. In this period of global warming, it should be obvious that scientists and government workers of all kinds, from Environmental Protection Agency researchers to sanitation workers, need to be confident that they can talk about what they know to be true without fear of retaliation.

Bargaining a strong academic freedom article that clearly applies to all contingents from the first day of their employment does not solve the problem of the self-censorship that of necessity flows from insecurity. However, a strong academic freedom article in the contract can provide a lodestar, a symbol for what should be. This is true both for combined units, where we would assume that the academic freedom article is the same for tenure-line as for contingent faculty, and for contingent-only units. This comparison can work to the advantage of contingent faculty if the union is actively supportive. Even in the case of a contingent-only bargaining unit, an academic freedom article can be important if a faculty member is courageous enough to fight an academic freedom issue and if the union is courageous enough to back them up.

It may not be obvious what kinds of issues are raised by the problem of academic freedom. Contingents, for example, often teach courses written by full-time tenured professors, and risk discipline if they change a reading, writing, or figuring assignment based on their own assessments of students' ability and need, or its cultural appropriateness. They may be expected to teach to a test written by a senior faculty person, whether or not the actual students they are teaching can be appropriately assessed by that test. In other words, contingents, lacking job security, are often placed in the position of conforming to the specifications of a curriculum prepared by someone else and setting aside their own skills and experience as teachers. This experience is likely to include more recent academic preparation or broader knowledge of teaching practice based on having taught at multiple other institutions.

Nor is it just a matter of being required, on pain of being fired or de-scheduled, to set aside their own knowledge of the subject matter and appropriate teaching strategies. In many classes, critically important but politically sensitive matters will come up. One could argue that classrooms are an ideal place for such matters to be brought into the light of day and discussed. But contingents without job security are vulnerable to the single student who complains to an administrator about the airing of a politically sensitive issue. An example of such an issue would be evolution, a topic that can show up not just in biology classes, but in others, like history, geography, composition or literature. Only too often, the administrator will capitulate to the student (who is being treated as a "customer" in situations like these) and will discipline the instructor.

A well-known example is the case of Douglas Giles, an instructor at Roosevelt University in Chicago, who in 2005 was teaching a course in comparative religions and, as part of the discussion of Judaism, mentioned Israel, Zionism, and the Palestinian side of the Palestine–Israel conflict. He was fired explicitly for this. His dean, who was not legally or contractually required to give reasons, actually sent him a termination letter and also took care to tell him on the phone exactly what her reasons were. Giles' union, the Roosevelt Adjunct Faculty Organization, IEA/NEA, took up the case and won, giving him a financial settlement and the offer of another class. Usually, administrators are not so foolish as to actually send a letter to a contingent who is being fired or even to explain orally what the issue was. Giles, luckily, had the dean's actual letter of termination in hand. He had also been able to type a word-for-word transcript of the Dean's telephone call, in which she specified her reasons.[1]

Other situations where the issue of academic freedom comes up are when a contingent "extramurally" writes, publishes, or is interviewed, appears in the media as an expert, public intellectual, or citizen, and says something that someone at the employing institution finds offensive or threatening.

There are good reasons not to try to be specific about what academic freedom actually is when bargaining for it at the table. Lists (such as the freedom to choose one's own assignments, tests, reading materials, topics, teaching methods, publishing extramurally, speaking in public) risk being treated as complete and exclusive rather than open-ended. This issue came up in CSU bargaining, as we will discuss when we talk about the CFA contract in the following chapter. The gold standard description is the current version of the 1940 AAUP Statement on Academic Freedom. This should be bargained straight into the contract with explicit application to all contingent faculty if possible.

The fundamental point of academic freedom is that teachers are responsible to know what they teach and to do it honestly. Attempts to interfere with this are violations of and attacks on academic freedom. The academic freedom of faculty to teach or research is the academic freedom of students to learn.

For our purposes, bargaining the concept of academic freedom is what is important. Putting academic freedom in a contract means putting it into a public document that is studied by both union members and administrators. The fact that it is in the contract makes it public. This is in contrast to the situation faced by contingents at institutions where there is either no union or no academic freedom article. In places like this, any discipline, de-scheduling, or firing will take place invisibly: the contingent who has offended someone will simply disappear. No announcement is made, no farewell party; the person is just gone. On top of the normal fear of a bad recommendation or blacklisting in the discipline and/or geographic region, there may be some kind of confidentiality agreement that binds the contingent to silence. Another consequence is the shadow of fear that these disappearances cast on others, leading to self-censorship. One contingent compared the heavy silence after such a dismissal to the silence that used to be the norm following a rape.

Administrators will usually hesitate to refuse to include an academic freedom article in a contract covering faculty in higher education. But once an item protecting the academic freedom of all faculty, including contingents, is written into a contract, then comes the organizing, convincing our unions and organizations to fight to defend it. There are cases where this has been done, but it requires an unusual amount of courage on the part of both the faculty member and the union. The violations of academic freedom that limit our ability to do our best work, along with job insecurity, remain the great negative markers of our contingency and precarity.

GOAL 7: THE COMMON GOOD

Beyond the table, we bargain to include those who are affected by what we win. The website of the organization Bargaining for the Common Good defines "common good" as any set of demands that benefit not just the bargaining unit but also the community as a whole.[2] Higher education teachers ("professors") have had in the past and could have in the future tremendous influence in the public sphere because of the respect that the public pays us. We can wield that influence not just individually, drawing on our prestige or expertise, but collectively through our organizations. When we consciously fight for the common good, our demands get more expansive. Although our demands still benefit ourselves, they cover more people and different kinds of workers. In the long run, these demands are more important than those we can reasonably hope to bargain under the current constraints on scope of bargaining and the limited power that we can exercise on that terrain.

The expression "bargaining for the common good" is a relatively new one flowing largely from the Chicago teachers' strike of 2012. However, the concept is as old as unionism and collective bargaining itself. Examples range from the demand for a day off to vote by early nineteenth-century unions after they won universal white male suffrage, through attempts by progressive unions such as the Packinghouse Workers, which bargained against racism during the Depression and after World War II by demanding integrated housing in Chicago, to early AFT attempts to bargain More Effective Schools (MES) in the 1960s and early 1970s.[3] As educators, we have the additional advantage of the truth of the slogan that our working conditions are our students' learning conditions, which gives us a definite leg up in building alliances, whatever the concept is called.

For instance, we can win bargaining rights for groups who were left out in the past. This happened in 2001 in Illinois where part-time community college faculty who were called "casual workers" because they taught less than 50 percent of load were brought in under the collective bargaining law by an amendment that the organized part-time faculty pushed for through the state unions, IEA and IFT, in a joint effort. The result was literally thousands of part-timers gaining union representation. In Washington State, due to the Washington Part-Time Faculty Association, contingents were brought under coverage through legislation and court cases for retirement, health insurance, and other issues. Issues like sick leave for all workers, which is critically important for contingents, became a state law in California with the support of the higher ed unions and contingents in them. Public transportation is another kind of issue that is particularly critical for contingents who may teach in many places with a metro area, and for our students. Housing is another, for the same reasons; there are housing fights going on all across the country, many focused particularly on teachers and students, at the time of writing.[4]

Recognizing that we contingents are now the majority of faculty also links to making sure that the public face of faculty reflects this reality and no longer hides behind the archaic image of the old, white, male, tenured professor. Dropping that image in our relations with the public may cost us some respect for the classic professorial model, but in exchange we can gain the much more valuable solidarity of the working-class majority who now, for the first time, can begin to see us as fellow workers, only in a different field.

By bargaining for the public good, we can win more than we could otherwise. Just as those of us who are white need to fight racial division and discrimination, not just on moral grounds but because it strengthens all of labor's struggle, likewise fighting for the common good, especially as public workers, allows us to build that broader movement that in turn can win what we cannot win by ourselves. These are points of synergy between the contingent faculty movement and other social movements.

We fight in the public arena for specific goals that cannot be achieved just through collective arguing. This public terrain is the terrain of legislatures at all levels: the courts, government agencies, and what has come to be called "civil society." These arenas use the tools of media relations, direct organizing and canvassing, political platforms, organizational and community alliances, support for ballot measures and politicians and civil society organizations of the sort called "nonprofits" in the US, as well as mass direct actions like the Women's March, Pride parades, immigrant workers rights and Mayday demonstrations, Occupy and the education workers' strike wave of 2018–19.

The most obvious of these is political campaigns and supporting candidates. We fight to get our unions and organizations to understand what has happened to academic employment, and to lead and support efforts to gain both adequate funding for higher education and equality among faculty. While avoiding being pitted against our colleagues and fellow unionists in the K–12 system, we fight for funding for higher ed in all its formats, but especially the most needy: the community colleges and adult education programs. These are where the majority of us—contingents—work, where the students with the greatest need study, and where the institutions have the fewest resources.

There are also examples of initiative campaigns that have relevance to contingent faculty, such as ones that propose progressive taxation, which lead to an increase in education funding, which in turn allows us to fight on a much more friendly terrain. While the initiative process exists in only a few states, the possibility exists in many more states to take an advisory role on issues of importance to us, like equal pay for equal work, revision of at-will employment, just-cause dismissal, and legislation to protect part-time and contingent workers. When we do this, we advance our own agenda and work with allies in the labor movement. Single-payer health care (Medicare for All) is an obvious example of this.

The first task is how to persuade your union to be interested in these issues, ones that are going to require resources, staff time, and expensive attorneys. Once the union is willing to put resources behind an issue, then comes the challenge of constructing the campaign or legal case. Court cases in particular should be prepared with special care. Our experience shows that these cases should be constructed collectively. Someone who pushes an individual case without collective consultation and committed backing can produce a court decision that sets a bad precedent that will harm other contingents far into the future.

GOAL 8: RIGHTS AND RESPECT IN OUR OWN ORGANIZATIONS

Unfortunately, though a great deal of progress has been made through the formation of contingent caucuses in the major unions and disciplinary organizations, a great deal more needs to be done to change the structure, the culture, and the programs of these organizations to more fully reflect the needs of the majority of the profession, namely contingents.

Primary, of course, is to animate these organizations and orient them toward the goal of abolishing contingency. Within that, we need to fight for access to the levers of power within the organizations that we do not control. Since we are the majority, this means democratizing these organizations to the greatest extent possible. Democracy in this context is a practice, not a set formula or list of rules, and certainly not just a matter of periodically voting on something.

Internally, democracy begins with organizing members. It may mean being able to elect our own representatives to offices or have them hired into staff positions. It may mean designated seats for historically oppressed and second-tier groups, like contingents but also others, like racial minorities, women, and the disabled. It might include equal votes despite our contingent status, representation in governing bodies (and access to them), and economic subsidies to participate in these organizations. To implement these, we need to push our organizations to provide adequate subsidies for travel to meetings and other organizational work. Likewise, within our own organizations of contingents, we need to practice a kind of transparent process and leadership from the base.

Programmatically, as we construct a more comprehensive, more detailed, constantly changing program for what contingent faculty need and how we should be fighting for it, we need at the same time to be fighting within our organizations to get them to adopt this program. This combination of organizing both within, yet distinct from, the official structures of one's organization is known as the Inside/Outside strategy (I/Os). We will go into further detail about the I/Os in Chapter 9 and 10 (especially Chapter 10) where we talk about the current network of contingent faculty movement organizations and their impact.

At this point, a list of the contents of a bargaining program for contingent faculty could be either two words, two paragraphs, or two pages. We will content ourselves with the following: equal compensation for equal work, and job security. These are the minimum conditions for academic freedom, the work of education workers. They are also of course the minimum conditions for workers in any job where doing the job right matters. The fight within our larger organizations will always be to win them to this program and then make sure we have our fingerprints on the details and implementation.

8
Beyond the Sausage-making:
A Close Look at the CFA-CSU Contract

The story of organizing among the Lecturers in the CFA starts with Lecturers knowing very little about their rights. From that base, they organized to gain resources within the union to further organize, protect and educate each other. Building on small successes, over time they became less intimidated and the landscape began to dramatically change. The key element was the increase in job security that came through steady grievance work in "careful consideration" and "equivalent assignment." As time passed, the average number of years Lecturers held onto their jobs and the average numbers of units each Lecturer taught increased, which made it possible for increasing numbers of Lecturers to focus their attention on their CSU jobs and their students there. The CFA was more able to put effort into improving salaries and benefits. By 1999, the Lecturers helped lead a revolution within the union. They could then move toward deep internal organizing. There came a turning point when the Lecturers re-defined the culture of the union—John Hess called it "the end of the tenured gaze." By the early 2000s, the CFA was able to self-organize enough to mount credible strike threats and win what is generally thought of as the best contract for contingent higher ed faculty in the US.[1] The contract that we will look at now is the tenth contract negotiated by the CFA, 2014–17, extended through June 2020.

Contracts evolve; each successive version builds on or concedes commitments defined in previous ones. What we see now is a contract that made big advances following the strike threat the union mounted in 2007 and built further on the political power developed between then and 2016. It is 140 pages long. We urge you to look at it in addition to reading this chapter. We will also look at the CFA Lecturers' Handbook, 2019–2020, which is available at the website www.calfac.org.[2] This handbook is written specifically for CFA Lecturers as a guide to the contract and to their rights in general. It originated in the early 1980s, patterned on the handbook of the UPC Lecturers' Committee from the 1970s, and has evolved since then. The existence of such a guide is in itself a product of the fight for rights inside the union.

Among our potential readers are activists on their own campuses who may not have much familiarity with any union contracts, their own or those of other unions, and who want to find out what is possible. To them, we offer two warnings. First, the CFA contract does not represent "all that is possible"—it does not end contingency and it does not provide equal pay for equal work. Nevertheless, it contains some real protections. Second, this contract is the result of arduous (a word used by the CFA) bargaining over many years, combined with a deliberate, conscious campaign to build the credible strike threat we talked about earlier. We will talk more about the concept of "credible strike threat" in our final chapter.

Finally, it is a fact that only by actually working under a contract can one report fully how it shapes someone's work life. Although we have talked with many people who have worked under these CFA contracts, we have actually worked under them ourselves only briefly. Before his death, we relied especially on John Hess for descriptions of his direct experience.

It will become clear right away that the organizing principle of our categories as described in Chapters 6 and 7 is not the same as the organizing principle of the contract. Our categories in Chapter 6 and 7 express our blue-sky vision of our role in relation to our students and the public. In this chapter, we look at how that blue-sky vision fares in a capitalist society when it goes through the fundamentally adversarial negotiations with a third party, the employer. Although the contract captures and to a certain extent fixes the power relationships of the moment, it is always a compromise and as such is a living document.

THE RIGHT TO COLLECTIVE BARGAINING

The rights to organize and to engage in collective bargaining were won in California when HEERA (Higher Education Employer-Employee Relations Act 1979) was passed in 1978. But these are rights that must be continuously fought for. Though HEERA allowed the CSU faculty to elect a system-wide collective bargaining agent, it did not prevent the bitter struggles that ensued while the faculty chose that representative. The fight, in other words, passed from the level of the state government to the level of the CSU system.

In the contract, the outcome of this struggle to date can be seen in the language of the definition of the bargaining unit, Article 1.1, "Recognition," which recognizes the CFA as the exclusive bargaining agent. What matters to us is, first, that the law would cover Lecturers; and second, that they would be in the same bargaining unit with tenure-line faculty. You can see this in the contract, but it's not in the recognition clause, where you might expect it. To find it, you have to turn to Article 2.13, where the definition of faculty bargaining unit employees is found.

Two further notes about the shape of the bargaining unit are important for contingent faculty: it explicitly *includes* department chairs and some other classifications that are directly involved in hiring, evaluating and assigning contingent faculty. This means that Lecturers may find themselves complaining about an action taken by their department chair, who is another union member. Nina Fendel explained in Chapter 2 how the CFA deals with this situation. Second, regarding the shape of the bargaining unit: one large category of contingent faculty, namely "Extension" and "Non-Credit" instructors, are *excluded* from the bargaining unit and in fact have never been organized in the CSU system, though they have at some other institutions, both community college and four-year.

What union recognition really means

We can see signs of the fight for the right to organize in Article 6, which cites the union's right to use CSU facilities. All campuses are required by the contract to provide office space for the CFA. This is important because it establishes the union as a legitimate presence on the campuses, embedding it in the physical space in a way that is both practical and symbolic. Practically, it creates acknowledged space on the ground for the union to bargain, organize and educate its members. For example, the CSU has to provide adequate facilities for CFA meetings depending on availability, access to phone and Internet systems, bulletin boards and intra-campus mail and campus mailboxes, while the CFA has to cover all costs of installation and operation. The CSU Human Resources Department is required to provide a monthly list of all newly appointed faculty unit employees. CFA representatives who are not campus employees (from other campuses or staff reps, for example) may come on campus, but an administrator must be notified of their visit.[3] Note that allowing the union to use campus facilities also supports the work of the union that focuses on "fighting for the public good."

Another important right is the right to union leave (Article 6.12). In the most recent contract, the CSU will allow leave up to 16 full-time positions per year to do union work, which will not "constitute a break in the faculty unit employee's continuous service for the purpose of salary adjustments, sick leave, vacation or seniority." Many unions negotiated leave for union activity, where the employer continues to pay the faculty member while they do union work. The CFA has chosen to negotiate to purchase faculty leave time. In other words, the union buys these positions from the employer using dues revenue and also controls who fills those positions, thus avoiding a potential conflict of interest.[4] The presumption is that union work has to be entrusted to people paid by the union. It also puts contingents on an equal level with tenure-line faculty because it is difficult, and sometimes impossible, to negotiate paid release time for Lecturers.

This may seem like a lot of union leave, but not if we consider that there are 23 campuses and about 29,000 faculty needing representation.[5] This is a substantial expense for the union in addition to the employment costs of its paid staff, but freeing some members to do the daily work of the union as part of their job is important for maintaining both the democratic character of the union and member involvement in it as an institution. For example, the most recent bargaining team consisted of 13 faculty, of whom five were Lecturers.

These details may seem like commonsense evidence of the university's desire to have a real partner to bargain with. However, these are things that the union has to propose; management will not propose them. Once in the contract, they must be defended. Many unions or informal gatherings of faculty have found themselves suddenly being told that meeting rooms are no longer available, or access to university communications services have been cut back or eliminated. Providing them had been "policy," and policy can change unilaterally. Faculty activism can easily be choked off by simply shutting down the means for people to get together.

We mention these items before going to the ones that most people will be interested in (compensation and job security), because the fight to bargain and organize has to be won before specific goals can be negotiated. Note also that under the contract these resources are accessible to Lecturers as well as tenure-line faculty.

COMPENSATION: WAGES

We have said that over time it became possible for someone working as a contingent in the CSU system to make enough money to lead a decent life—if they are assigned enough classes, of course, and if they live in a part of California where there is affordable housing. How much money is that, exactly? This is all public information, available on the Internet. This point is worth emphasizing because in some systems, salary information is considered confidential and employees are forbidden to share it with other employees.

The contract articles related to compensation for Lecturers are found in Articles 12 and 31. Article 31 refers us to Appendix C. If we go to Appendix C and look at the titles "Lecturer—Academic Year, and "Instructional Faculty—Academic Year" (which means the tenure-line faculty), we can see that both categories appear on a table of records that includes 1,482 total records. Pages 48–49 and 50 of the *CFA Lecturers' Handbook* provide the necessary interpretation.

Salaries for contingent faculty in the CSU system are expressed in terms of "Ranges."[6] When you are hired, your appointment letter will tell you what your range is. It is based primarily on your qualifications (degrees, experience, etc.). In Table 8.1, the first column, Range, is related to the title (column two, Class title) that a hire will hold upon being employed; Lecturers could be employed in Range A, B, C, or D. The lowest range for anyone with a terminal degree (a PhD or, in the case of Art and English, MFA) is Range B. However, about half of Lecturers in 2019 had MAs, not PhDs. The third column is the minimum salary per month for a full-time employee in that range teaching a full load of 15 units or five classes. The fourth column is the maximum.

For example, to get an annual salary for a Lecturer hired into Range B, the Assistant Professor range, who is teaching a full load of five classes at the maximum for the range for two semesters, multiply $11,197 by 12 months to get $134,364 per year.

More explanation: for tenure-line faculty, a full load is four 3-credit classes per semester. A full load for Lecturers is defined as five 3-credit classes per semester, or 30 units per year. Lecturers, however, can teach up to a full load and even up to 120 percent of a full load. It is worth noting that the ranges—

Table 8.1 Job position titles and pay in the CSU system[7]

Range	Class title	Min. full-time base Salary (Per Month) for teaching 5 classes or 15 units. For annual salary, multiply by 12.	Max. full-time base Salary (Per Month) for teaching 5 classes or 15 units.
Range A	Lecturer, Academic Year This is for someone who is hired into a department and does not have the same terminal degree as is standard for that department.	$4,229 per month Per 3-credit class: $5,074.80	$5,654 per 3-credit class: $6,784.80
Range B Assistant Professor	Lecturer, Academic year Contractually, any hire with the same terminal degree as standard for that department must be hired into at least Range B.	$5,406 per month Per 3-credit class: $6487.20	$11,197 Per 3-credit class: $13,436
Range C Associate Professor	Lecturer, Academic year	$5,779 per month per 3-credit class $6,934.80	$12,296 per 3-credit class: $14,755.20
Range D Full Professor	Lecturer, Academic year	$7,276 per month per 3-credit class: $8,731.20	$12,880 per 3-credit class: $15,456

the high and the low—are wide. Placement within a range is discretionary at the department level, but must be approved at the dean's level.

It is important that CSU contingent pay is on the same salary schedule as tenure-line faculty. This is unusual. A Lecturer appears to be paid the same amount as a tenure-line faculty member, considering the difference in load and allowing for placement within a range. Contingents also get service salary increases (SSIs) after they have taught a certain period of time in a single department (see p. 51). In addition, a dedicated fund for implementing contractually negotiated salary increases for non-tenure-line faculty was negotiated, as a step toward equalizing faculty pay.

These figures will look high to contingents in other parts of the United States, who may be working for as little as $1,000 per class. Recent reports put typical contingent compensation in the area of about $3,000 per class. On the other hand, they make the demands of Professional Staff Congress (PSC) at the City University of New York (CUNY) for $7,000 per 3-credit class look modest.

Benefits and access to rights

In Chapter 7, we also listed the non-salary economic factors that we fight for: access to healthcare insurance, retirement plans, and a commitment to our rights to workers compensation, unemployment benefits and disability benefits.

In the 2002 contract, as a result of both state-level legislation initiated by the CFA and bargaining, Lecturers who teach 40 percent of a full-time load got employer-paid healthcare benefits (health, vision, dental coverage, and life and disability insurance) identical to tenure-line faculty benefits. This lowered the benefit eligibility line from 50 percent to 40 percent, and affected thousands of people. If a teacher has non-CSU coverage from a covered spouse and they do not get the full value of the contractual benefit, there is a "Flex-cash" reimbursement for the difference. Lecturers have leave rights and full-time Lecturers can get a sabbatical. They have pro rata sick leave and can donate their sick leave to colleagues who need it. They are covered by the tuition waiver once they obtain a three-year appointment. This waiver also covers family members. If someone drops below 40 percent of a full-time load, they can keep their health benefits if they pay for them. They also get full access to recreational facilities, travel expenses for academic professional work, research grants and personal holidays. This represents a level of contractual equality that is extremely rare.

We have noted that it is possible to negotiate who in the employer's office responds to an unemployment claim and what they say. The CSU administration was, as far back as 1979, directing campus administrators to "control costs," coming from Lecturers' claims, which represented 50 percent of the CSU system's unemployment insurance costs. They did this in language that would be immediately interpreted as a directive to intimidate people in exit interviews and to file gratuitous appeals. The CFA helped get an Assemblyman elected who sponsored a bill, AB 2412, co-sponsored by the CFA and other faculty organizations, that said, "Any employer who gives false information to the Employment Development Department (EDD) is subject to a fine in the value of ten times the value of the unemployment benefit." The CFA then bargained this into the contract.

This further supports the value of the well-designed and far-ranging 1989 case from AFT 2121 at City College of San Francisco known as *Cervisi* (after the first plaintiff), which was appealed to California State Court of Appeals. The Court found that non-tenure-line faculty do not in fact have "reasonable assurance of re-employment" and are therefore eligible for benefits.[8]

JOB SECURITY

The CSU-CFA contract shows how you can bargain a seniority system without ever using the word "seniority." For most contingents, although salary matters, job security is even more important. Lecturers in the CSUs have a degree of real job security. Nevertheless, they are hired on temporary contracts, not permanent tenure-line contracts. "Temporary" contracts can be for one quarter, one semester, one year, or three years. The rights to three-year appointments and existing Lecturer preference for new work were won in the 2002 contract, in the first round of new bargaining after the Revolution and are in Articles 12.29 and 15.28, the latter article stating:

A three year appointment *shall* be issued if the temporary faculty unit employee is determined by the appropriate administrator to have performed in a satisfactory manner in carrying out the duties of his/her position. The determination of the appropriate administrator shall be based on the contents of the Personnel Action File and any material generated for use in any evaluation cycle.[emphasis added]

"Shall" means that it has to happen; it is not up to a whim of the administrator or department chair making assignments. Getting your assignment is not a privilege or a piece of good luck. If you have been teaching a specific assignment (such as a course) for two semesters, or three quarters, consecutively, you should expect to count on getting that course to teach again for a full year if it is offered. In addition, according to Article 12.9, you should expect the same or a higher salary placement. After six or more years of prior consecutive service, you should expect a three-year appointment (Article

12.12). The entire six years has to have been worked on a single campus in a single department. The only time the CSU can hire a new Lecturer is when all the incumbent three-year and one-year Lecturers are hired full-time (this is not the same as tenure line), unless there is a new course that requires someone with different expertise. This is how a part-time Lecturer can spend thirty-plus years in continued employment in the CSU system. Barring any changes in tenure-line hiring or a budget meltdown, continuity is the expectation.[9]

It is the fact that although these are contractual rights that one can expect to exercise, classes are still awarded on the basis of enrollment and funding and are therefore contingent. If there is not enough available work, a Lecturer's right to that assignment continues for three years on a public list. If a class is cancelled for budget or enrollment reasons, you would either be paid for the portion of the term worked, or, if the class has met three times, for the rest of the semester. The contract provides administrative flexibility, but if "no work exists to support the initial or subsequent three-year appointment," the faculty member shall be placed on a list for re-employment.

Approximately half of summer assignments go to tenure-line faculty; after that, there is a list of Lecturers who are offered the remaining classes. The CFA has to watchdog these lists to make sure that the agreements about the order in which assignments are made are respected. Even though the word "seniority" is never mentioned, it is in practice respected in the cascade of criteria for who gets what assignment; this section of the contract is three pages long. Errors or the practice of favoritism in job assignments that violate the negotiated sequence can be grieved, as can disparate treatment in general.

Just as a reminder: our definition of job security for faculty, also known as tenure, is simply "just cause dismissal," as in other kinds of work.

"Careful consideration" as a way to get some job security for Lecturers

Article 12.7 contains the key words "careful consideration." These words first appeared in the earliest 1983 contract. It applies to retention in Lecturer positions and access to new work in Lecturer positions. The contract says, "the employee's previous periodic evaluations and his/her application shall receive *careful consideration*" (our emphasis). These words are critically important. What exactly does "careful consideration" mean? How do you

show that an administrator who has made an assignment has either performed or failed to perform "careful consideration"?

Nina Fendel described the two mass layoffs in the early 1990s. In the
first layoff, hundreds of Lecturers were losing their jobs across the CSU
system. The administration position was that there was not enough money
to support those classes. The union did two things: one was to research
the actual budget, with the result that they located pots of money, much of
which was buried in foundations. Each campus had a foundation, which
collected a percentage of funds that came in through grants on the basis of a
charge for overheads. This money then became discretionary for the administration to use. The union had made contacts among workers in the finance
and budget departments, who cooperated happily in this research.

Once these pots of money had been identified, the union took up the cause
of the Lecturers whose jobs were being cut. They grieved that the Lecturers'
loss of assignments was not given "careful consideration." One marker for
whether or not a Lecturer had been given careful consideration was whether
that Lecturer's personnel file had been signed out of the personnel office
where the official files were kept. (Files that may or may not have been kept
in a department chair's office were not the official file.) If no one had even
signed out the official file, it was clear that that Lecturer had lost their job
without careful consideration. The union put a pile of grievance forms on
a chair outside the union office on each campus and developed what Nina
Fendel called "an assembly line" for processing them. In this way, hundreds
of Lecturers got their jobs back.

The second wave of layoffs happened in Spring 1992, and this time there
was no secret pot of money with which to fund Lecturer jobs, which were,
after all, *contingent* on enrollment and funding. John Hess was among those
who were laid off at this time.

Another key contract phrase, bargained at the same time in 1983, is
"similar assignment." Not only does the administration have to perform
"careful consideration," but the Lecturer must be offered a *similar assignment*,
meaning the same load, or no cut in pay (see p. 12 of the CFA Handbook).
One way to talk about this is "share the wealth, not the poverty." This contrasts with the share-the-poverty system, where the classes are assigned one
at a time to everyone on the seniority list before anyone gets a second class.

Both "careful consideration" and "similar assignment" became the subject
of grievances and arbitrations early in the development of the CSU-CFA

contract and forced administrators to be able to justify their action. Arbitration decisions on this are summarized on pp. 15–16 of the CFA Handbook, underscoring the kinds of actions necessary to make contract language real. The point is that the words in the contract can be made more significant and real to the lives of the members based on the grievances that are filed, the arbitrations that are done, the awards that are achieved and the union's willingness to enforce and build on existing language in the contract through these processes over a period of years.

However, the union had to answer this question: "Careful consideration of *what*?" This in turn led to winning an evaluation process that is hedged around with protections for the faculty member.

Evaluations: A portfolio process

The right of a Lecturer to careful consideration is predicated on having had a proper evaluation. The following gives only a sketch of the full process. The contract article covering evaluations is Article 15. Evaluations are mandatory after the second semester (though a person can request one in the first semester). They are a portfolio process; the judgment is based on documents that are assembled into a portfolio called a "Working Personnel Action File." The faculty member identifies the materials that they want to have included for consideration in this portfolio. The portfolio, not the subjective assessment of behavior, attitude, family connections, eagerness, or future potential of the faculty member, is what is being evaluated.

Students, administrators, and other faculty employees may contribute materials to this portfolio, but it is never just students. Many colleges and universities gather only student evaluations and give them significant weight. Evaluations of faculty members who are eligible for a three-year appointment must include student opinion surveys and recommendations from a peer review committee and administrator reviews (Article 15.28). Faculty members must also prepare a curriculum review, demonstrating how they have remained active, productive and current in their discipline and pedagogy. Classroom visits must be arranged five days in advance and there must be consultation between the faculty member and the visitor. The faculty member is allowed to see the written responses to the portfolio: "At all levels of review, faculty unit employees shall be given a copy of the recommendation and the written reasons therefore" (Article 15.15). The

CFA website contains announcements of discussion sessions to help faculty prepare rebuttals to items that come up in this evaluation process.

The essence of this process is the focus on evaluating the contents of the Working Personnel Action File, and right of the faculty member to place materials in that file, to have access to it, and to monitor and rebut things others place in it. Because this process is in the contract, a violation of it can be grieved, which is given a ten-day window to give the Lecturer time to submit materials to the Dean and then the Chair of the Department. However, in order to grieve an evaluation, the faculty member must show harm, namely not being re-hired or given a lower load.

The assumption is that what has gone into that file has been placed there through a fair process and therefore should reflect an accurate picture of the faculty member's performance. This is "careful consideration." According to Item 15.12.c:

> Personnel recommendations or decisions relating to retention, tenure, or promotion or any other personnel action shall be based on the Personnel Action File. Should the President make a personnel decision on any basis not directly related to the professional qualifications, work performance, or personal attributes of the individual faculty member in question, those reasons shall be reduced to writing and entered into the Personnel Action File and shall be immediately provided to the faculty member. For the purposes of this section, course assignments shall not be considered personnel actions. However, course assignments shall not be punitive in nature.

However, only tenured faculty and academic administrators may "engage in deliberations and make recommendations to the President regarding the evaluation of a faculty unit employee" [15.2]. This is in our opinion a weakness in the contract, and an argument could also probably be made that this setup is a burden on tenure-line faculty.

ACADEMIC FREEDOM

Although compensation and job security are the issues that most contingents will look for when reading this chapter, we are also sure that academic freedom is high on their list of concerns. Academic freedom is an ideal and is described in the AAUP 1940 *Statement of Principles on Academic Freedom and Tenure*: "The common good depends upon the free search for truth and its free exposition. Academic freedom is essential to these purposes and applies to both teaching and research."

However, in practice, academic freedom depends on very concrete supports, including channels of communication, scheduling of meetings, representation at those meetings, professional development opportunities, and intellectual property rights, among others. Concrete actions that academics must be able to carry out freely include choosing assignments, managing research projects, awarding grades, advising students and otherwise exercising professional judgment. The words "academic freedom" are mentioned in the preamble to the contract: "The CSU shall support the pursuit of excellence and academic freedom in teaching, research, and learning through the free exchange of ideas among the faculty, students and staff."

But freedom is balanced by responsibility:

> The parties recognize that quality education requires an atmosphere of academic freedom and academic responsibility. The parties acknowledge and encourage the continuation of academic freedom while recognizing that the concept of academic freedom is accompanied by a corresponding concept of responsibility to the University and its students.

Otherwise, the words "academic freedom" are not mentioned specifically in the CSU-CFA contract. Instead, we find the issue of academic freedom coming up in the CFA Chapter By-Laws. One of the purposes of the CFA chapter (in this case, from the Los Angeles CFA chapter) is to "seek to obtain explicit guarantees of academic freedom and tenure." This is to be done through "orderly and clear procedures" (such as through the grievance process), or else through some other unnamed process. This entire subject applies, in this context, equally to Lecturers and tenure-line faculty, but in practice it is much harder to enforce these provisions on behalf of Lecturers.

However, during bargaining in May 2019, the CFA attempted to incorporate some language from the AAUP Statement on Principles of Academic Freedom and Tenure into the new contract. This was in response to the increasingly partisan political climate, which raised concerns among faculty. The CSU management responded by asking to add a list of prohibitions on faculty speech and expression in an appendix to the contract, including subjecting faculty to disciplinary action if they violated those prohibitions. The CFA team then decided to leave the current contract language in place.

Intellectual property rights

The CSU negotiators made a similar response in May 2019 when bargaining over the question of intellectual property rights. Intellectual property is of special interest to contingents because many of us have second jobs. We create classes, write assignments, design reading lists, etc. We may use the same materials in more than one institution or in other contexts. When teaching online, we may have to write an entire course out in advance, including all our lectures, so it can be uploaded to a website owned by the institution where we teach. Some of our intellectual property has significant financial value (books, artwork, creative work generally, including technology or scientific work). Contingents sometimes ask who owns the course they have designed and re-worked over many semesters: does the university own it, or do they own it, and can they take it to another university or publish a book using it?

The CSU-CFA contract (Article 39.3) says: "Faculty bargaining unit employees may use for non-CSU purposes materials created by them without extraordinary University support, if in the past the CSU has never disputed the use of such material … for non-CSU purposes." This includes works used in online or hybrid instruction, a particularly difficult type of course to prepare and one that a contingent would be particularly concerned about giving away all rights to. The catch is in the "extraordinary University support," which might include special funding for the development of some project. Also, the contract says that neither party waives its rights to assert ownership over materials created *without* extraordinary support (Article 39.2); Article 39.4 cautiously warns that neither party waives the rights to assert ownership over materials created in "new and emerging media of expression, regardless of whether either party has ever asserted a right of ownership in the past." This sounds like looking ahead into an increasingly adversarial but automated future.

And indeed, in the Spring 2019 bargaining phase, the CSU team wanted "intellectual property [to] default or be licensed indefinitely to the CSU with no or minimal financial and institutional support," whether or not its development had received extraordinary university support. On this issue also, the CFA team decided to let the contract continue without changes.

Two of the goals mentioned in Chapter 7 under "Fighting for the public good" show up in the CSU contract where it talks about the quality of education, the role of the university in the community, and equal treatment for students and faculty especially of diverse backgrounds. This issue shows up under workload, as well, since class size and faculty workload strongly impact the quality of the education that students and taxpayers are paying for.

Quality of education, role in the community, and equal treatment

Article 16 of the CFA-CSU contract is the nondiscrimination clause and prohibits discrimination against employees on the basis of 16 different characteristics, including gender identity, gender expression, genetic information and medical condition, in addition to the more traditional characteristics like race, color, religion, ancestry, etc. It establishes a statewide joint committee of CFA representatives and administrators who are charged with gathering and exchanging information and making reports and recommendations. They may also "make recommendations regarding efforts to facilitate the instruction of diverse student populations." This means that this committee's function is not just corrective of past abuses but can also shape the future. Looking back to Chapter 1, where we talked about the concerns of the students on strike at San Francisco State in 1968, which led to the first School of Ethnic Studies in the US, one could say that the contract has come a long way.

Related to Article 16 is Article 20.37, "Research, Scholarly and Creative Activities." A pool of $1.3 million per year is set aside to support activities focused on "underserved, first-generation, and/or underrepresented students." This pool became accessible to Lecturers in 2016. Such activities would include re-design of curriculum, service to the community, and taking on courses where "increases to enrollment have demonstrably increased workload" as well as other activities. This is partly in response to the widely received problem of faculty members of underrepresented groups being tasked with extraordinary levels of work to support and mentor students and colleagues. Members of these same underrepresented groups are also often asked to provide diversity on multiple institutional committees.

Workload

Contingents often mention the axiom, "Our working conditions are our students' learning conditions." We want to see some protections for good working conditions in the contract. For this we can look under Article 20, "Workload." The item begins with a generous list of the duties of all faculty, not distinguishing between contingent and tenure-line faculty. These duties include preparation for class, research, scholarship and creative activity "which contribute to their currency" (Article 20.1.d) such as participation in conferences and seminars, and consulting with students and colleagues. As far as class size goes, the contract states: "Article 20.3.a. Members of the bargaining unit shall not be required to teach an excessive number of contact hours, assume an excessive student load, or be assigned an unreasonable workload or schedule." The contract, however, does not define what is "excessive" or "unreasonable." Instead, it includes this guideline: "Article 20.2.a. The composition of professional duties and responsibilities of individual faculty cannot be restricted to a fixed amount of time, and will be determined by the appropriate administrator after consultation with the department and/or the individual faculty member."

The person who makes this determination is obviously the administrator, but the contract also provides a list of concerns that must be considered in determining workload, including the number of students who want to take classes in that area, the level of support provided by the program, and past practices of the university, all of which provide a basis for a discussion between the individual instructor, the CFA representative, and the administrator. This also provides a basis for a grievance if these are violated.

The presence in the contract of items that go beyond traditional job issues and take up issues of the quality of education, especially access to it by members of traditional minority populations, indicates how much the faculty union views its contract as being one that is really a three-way agreement—between the union, the administrators of the university system, and the public. "The public" here is not the elite 1 percent alumni, whose donations would enable the university to hire star faculty, promote competitive sports teams, or erect fine buildings; it is the actual public of California, which is now 51 percent "minority" and increasingly in the lower tiers of the 99 percent.

The last item in our Chapter 6 list of what we fight for is rights within our own organization. The question of rights within the union is not taken up within the contract. They are found in the union by-laws and constitution and are detailed on p. 43 of the CFA Lecturers' Handbook. The administration can have no power to interfere with those rights, either to support or undermine them. However, the contract gives them a concrete reality that would not exist if they just came from provisions established within the CFA for Lecturers. Through rights within the CFA, contingents have rights that impact negotiations and agreements with the university.

To look for those rights, one turns to the CFA Lecturers' Handbook. For example, on p. 7:

> Lecturers ... have become an important constituency in CFA, which guarantees special representation to Lecturer members in union affairs and governance. For example, the CFA by-laws call for specific Lecturer representation at the CFA Assembly (which is the statewide governing body and direct representative of members), among the CFA Officers (of which two are the AVPs of Lecturers), on the Board of Directors (BOD) (which oversees the governance and carries out the politics of CFA), and on all statewide committees) such as Political Action/Legislation, Representation, and Membership and Organizing.

In 2020, five Lecturers sat on the CFA Bargaining Team, four on the Contract Development/Bargaining Strategy Committee and also chaired other committees. Many of these are "designated" seats—that is, they must be occupied by Lecturers. Six Lecturers are required on the Board of Directors, four on the Contract and all Statewide Committees must have Lecturer representation. The capacity of the CFA to field this many faculty, whether contingents or tenure-line faculty, to do union work is related to Article 6.12, "Union Leave," noted in the contract above. Contingents can get paid (by the union) to do union work and be confident that they can return to teaching without losing seniority or benefits during their leave. But just as the contract is the result of negotiations and struggle over the years, so these conditions within the CFA have also been the result of negotiations and struggle, going back to the 1960s.

The CFA Lecturers' Handbook then goes on to summarize the situation of contingents in the CSU overall, followed by a history of advances in protections for Lecturers since the 1983 contract. They provide a "quick guide" to Lecturer provisions in the contract under four categories: "Getting and Keeping the Work," "Getting a Fair Salary," "Getting Benefits," and "Equal Access/Recognition as a Faculty Member in the System" (p. 12). These correspond to the categories we listed under "What We Fight For" in Chapters 6 and 7.

An outsider looking at this contract can hardly begin to imagine the work invested in tooling and re-tooling the language of these articles. What is also invisible, unless you know something about working in higher education, is the political muscle that has had to be developed to even get to the table to negotiate these articles.

PART IV

THE DIFFICULTY OF THINKING
STRATEGICALLY

9
Strategies Emerging From Practice

In this chapter, we move to higher-level strategic questions. The contingent faculty movement is at a point where a discussion of strategy on a national level is both needed and possible. The strategy we have observed thus far, which we call the Inside/Outside strategy, was arrived at intuitively over the years and to a great extent unconsciously, in the search for effective practice. An example is the unexpected realization John Hess had, having brought the Ruckus Society to an organizing session for CSU contingents, when he notices the tenure-line faculty leadership watching the session with excitement, and realizes, "They have nothing to teach us," upending that relationship between tenure-line and contingents.

Outside of the contingent faculty movement, this strategy of self-organizing a subaltern group for power in a way which is not part of the architecture of the larger organization, has precedents which we discuss in this chapter, and which were theorized to varying degrees by the people involved. We also want to look at some of the building blocks that laid the basis for our strategy. One of these is in the work of Antonio Gramsci. As far as we can tell, he never used the term "inside/outside," but today almost all theorists of that strategy look back to his *Prison Notebooks* for the roots of their own thinking and practice. Among the most important of these concepts is that of ideological and cultural hegemony, followed by the methods to be used by subaltern groups who wish to subvert that hegemony. Gramsci of course recognized, as a good communist, that he was drawing from Marx who said that in any era, the ruling ideas are the ideas of the ruling class.

Another concept that Gramsci did not invent but did a lot to popularize was the idea of workers' councils, growing out of the *soviets* of the Bolshevik revolution. Gramsci saw this organizational expression as different from either trade unions or political parties, both of which he also participated in and supported. He viewed workers' councils as the seed of working-class control in a future socialist society. The councils would be both broader than the workplace and more rooted in the regular daily life of workers than political parties. There are parallels here with the independent formations that the contingent faculty movement has attempted to create over the years of its existence, that speak to the direct conditions of us as workers and also to the content of our work in education and its contribution to the public good.

THE DIFFICULTY OF THINKING STRATEGICALLY

The 2004 Coalition of Contingent Academic Labor (COCAL) conference in Chicago took place over the objections of both the NEA and AFT leaderships. They viewed it as a distraction, with no potential useful outcomes. Despite this, the conference had substantial representation from Mexico and Canada, including Quebec, and between 200 and 250 people attended. The existence of an independent national network had become visible and confirmed.

The planners of the conference decided to ask participants to submit papers in which they would attempt to analyze, propose and discuss strategy for the contingent faculty movement. This had not been done before. In fact, at that time very little strategic analysis had been done. Most of what had been written from the point of view of contingents consisted of memoirs, individual case studies, and articles exposing the "plight" of contingents.

As the papers for the strategy panels came in, it became obvious that thinking strategically about one's own situation was difficult. Simply saying, "I am in a bad situation," was not enough. What was required was something more than telling one's own story, even if as a case study of a local struggle. Nor was quoting the national or local statistics about how many faculty worked as contingents. Between the personal stories and the national statistics was needed an explanation that showed a grasp of the purpose of the industry of which contingents are a part and how that industry is itself a terrain of struggle. A strategy for contingents required abstracting generalizations and then testing and theorizing those generalizations against the broader political economy in which we operate.

However, at the 2004 COCAL conference, the discussion of strategy was very limited. Although enough papers were submitted, they did not grapple with the key issues theoretically or strategically. Time and again, people fell back on telling their own stories. Participants found it difficult to maintain an ongoing, consistent discussion that integrated their own experiences and ideas with those of other people in the far-flung sections of the movement in a productive way. This demonstrated the lack of theoretical experience among contingents at the conference.

The tool that is necessary in order to rise above one's own experiences and compare it with that of others is theory. Theory consists of generalizations, based on definitions that are agreed-upon and understood tested against evidence. A theory is not a law of nature but instead something that has to be adjusted over and over again, while being used. Furthermore, a theory always exists in the context of other competing theories.

While the conference was considered a success overall, it did not produce the theoretical thinking and materials that the organizers had hoped for. Like most social movements in the United States, one reason for this deficiency in theoretical thinking among activists is that people are not trained to think theoretically, critique theory, or even respect the idea of theory. Movements, perhaps especially in the labor movement, have repeatedly been limited and constrained by this widespread anti-theoretical tendency in our culture and education. This is especially linked to anti-communism, and the taboo on Marxist theory in particular.

RECOGNIZING THE THEORY WE DREW ON IN PRACTICE: THE I/OS

Today in 2021 both the discussion and the practice have advanced. The fact that the SEIU has moved to a metro strategy,[1] organizing a regional workforce rather than just the employees of one specific institution, is evidence of this. So is the attempt by New Faculty Majority to establish a stable independent organization that could carry an agenda from year to year, a role that COCAL, which has no regular staff, could not fill. However, we are not yet at a point where we can say that we have a tested general strategy for how to get what we fight for in higher education.

Forty years of past practice indicate that an Inside /Outside strategy (I/Os) is what the contingent faculty movement has followed, consciously or not, to get to where we are now. An I/Os is when a sub-group organizes itself as an independent base of power within a more powerful group in order to create a safe space where they can have a significant impact on the more powerful group. "Outside" is a political term here, not an organizational term. It means organizing ourselves *inside* whatever overarching body (like a union) we are part of, *outside* of (in the sense of different from) the existing official structures of that body.

An early description of this strategy can be found in a letter written by John Hess to the then-chair of the UPC Lecturers' Committee in June 1981. This would have been in the middle of the fight over whether contingents would be included in a single bargaining unit with tenure-line faculty. He may have been following the advice given to Lecturers by the members of the UPC Affirmative Action Committee six years earlier, during the early years of the UPC/CFA competition, who had experience self-organizing as a safe space for people of color within the CSU faculty. John Hess explained why Lecturers should be part of the same bargaining unit with tenure-line faculty, but also why they should organize themselves as a distinct group within that bargaining unit:

> The history of the American Labor movement is the sacrifice of the unskilled by the skilled, of the lower paid by the higher paid. The people with the power in the union are able to get what they want at the expense of what the less powerful want. It seems that two things tend to happen in our situation: if full- and part-timers are in separate units or even unions, the administration is able to whipsaw them—divide and conquer. If they are in the same unit/union, the Lecturers disappear, [they] don't participate in the union and are generally ignored. The only way we can avoid this is if we have a strong Lecturers' group, an independent power base that is able to function with a certain cohesiveness within the union to keep people honest and to defend our interests.

Thus going back to the 1970s, parallel with the increase in the numbers of casualized faculty, contingents were already holding meetings together, forming caucuses and committees, creating communication tools dedicated to their concerns and otherwise organizing themselves, "outside" of their existing union structures and workforces (but not seceding from them), in order to also have power within—"inside"—their unions and workforces.

In the 1970s, 1980s and early 1990s, this creation of independent safe spaces was locally focused. Organizing, recognition and bargaining take place at an institutional level, and people have to fight at the level where these happen, so that is where the focus was at that time. These were caucuses and committees as well as just groups of friends talking with each other, maybe around planning to write something. They were rarely in touch with each other at a statewide level and hardly at all nationally. People wrote books, book chapters and articles that reached a national audience, but there were no easy communication tools that could rapidly organize a meeting across time zones. Conferences took months to plan and were expensive to put on and to attend. So there was no national contingent conversation about a strategy. None of the major faculty unions were interested in leading such a broad strategic discourse. Therefore local contingent activists found themselves coming up with and engaging in local versions of the Inside/Outside strategy, because they felt they had no alternative. These I/Os versions were often parallel to each other, but they were uncoordinated and usually unconscious of the strategic implications of what they were doing. They also didn't have a name, even though some activists, like Hess, were themselves socialists with some general theoretical training.

Importantly, local contingent activists came to this I/O strategy because they were not just facing hostile management negotiators across the table, they were also facing resistance from within their unions at the local, state and national level. These more powerful bodies *did* coordinate with each other. They did not merely fail to support the contingent activists, they often actually blocked them. The national unions (the NEA, AFT and AAUP) were all resistant, even hostile, to paying attention to contingents. Many state and local unions went out of their way to prevent contingents from gaining power. Contingents needed not only a place to organize to confront their employers; they needed one to deal with the way they were treated by their own national unions.

There are some successes from that period where contingents self-organized and carried out an I/O strategy. In addition to the limited successes of the CFA Lecturers, there was the California Association of Part-Time Instructors (CAPI), a statewide community college group in California in the 1970s, that generated local groupings and a statewide committee inside the CFT, along with a major lawsuit. CUNY Adjuncts Unite was a group that started in the 1970s, which spoke inside and outside the union and even ran a decertification campaign at one point because they did not feel they were being represented. They met severe repression, but succeeded in changing the terrain for the next struggle, which substantially altered the leadership of the local union.[2] That new leadership ultimately built for a credible strike threat and in their 2019 bargaining won pay of over $5,000 per class for adjuncts. Their story parallels the CFA story in many ways.

However, there was really no place to pull this experience together at a national level—no "safe space" for outside organizing to take place—until the late 1990s. Nor was there a means: this was in the early days of email and Internet forums. In other words, there was no national "outside."

THE BEGINNINGS OF A NATIONAL MOVEMENT

The key event that changed the national contingent movement landscape took place in conjunction with the December 1996 Modern Language Association Convention. The MLA is not a union, nor is it part of the labor movement. Instead, it is a professional disciplinary organization. The Graduate Caucus and the Radical Caucus of the MLA called for a National Congress of Adjunct, Part-time, Graduate Student, Contingent and Non-Tenure-Track Faculty in Washington, DC. It was more successful than anyone expected and led to another congress the next year in New York, largely organized by Vinnie Tirelli, a contingent in political science at CUNY, and Eric Marshall, also a CUNY contingent. At that time, the list-serve known as adj-l (for adjunct list serve) was started. It survives today and is still moderated by Tirelli.

In 1997, at the end of the New York congress, the name COCAL (Coalition of Contingent Academic Labor) was adopted. The first meeting held under the banner of COCAL took place at the University of Massachusetts (UMass) in Boston in 1999. It was independent of, but still had support from, the Graduate Caucus and the Radical Caucus of the MLA. The main organizers were Gary Zabel, a contingent in philosophy at UMass Boston; Rich Moser, who was working as an organizer for AAUP at the time, and Jon Bekken, a tenure-line professor of journalism who was also editor of the IWW newspaper, *Industrial Worker*. By that time, the national unions had published statements bemoaning the "plight" of contingents, but had not taken concrete steps to change conditions.

COCAL in 1999 was followed by more conferences. It now takes place every two years, rotating among the US, Canada, and Mexico. At about the same time, in the late 1990s, the North American Alliance for Fair Employment (NAFFE) came into being with its Campus Action Group, which accessed foundation funding for face-to-face meetings of contingent faculty leaders. The funders included the Ford Foundation, the Rockefeller Foundation and the French-American Foundation.

The "safe space" was beginning to take shape. Now, with a certain expectation of continuity, some visible leaders and some communications tools, it was possible to debate and practice an I/O strategy on a national level, reproducing what were both the conscious and unconscious lessons of the previous thirty years of activity at a local and state level, especially in those places where casualization had been earliest and most advanced, such as New York, Massachusetts, Washington State and California.

From 2004 onward, the continuing capacity to successfully develop this national network revealed an existing independent power base for contingent faculty that had not been apparent before. This power base in turn drew support from the national unions. Thus progress with the outside part of the strategy made it possible to move to the inside of these organizations on a more effective level.

PRECEDENTS FOR THE INSIDE/OUTSIDE STRATEGY

One precedent for the I/Os was the Teamsters for a Democratic Union (TDU) in the 1970s. John had first-hand knowledge of this reform effort because Gail Sullivan was active in the TDU and worked at the national level when they were in power in the 1990s. Another precedent was the Miners for Democracy (MFD) in the United Mineworkers, active during the 1970s. And yet another precedent, the Dodge Revolutionary Union Movement (DRUM) in the 1960s, was specifically formed around issues of racial discrimination by both the company, Chrysler, and the union, the UAW. These were all familiar to the contingent faculty leaders in the 1990s. All of these can be seen as groups deploying the I/Os, and they share a common goal of equality and structural reform in the face of a hardening political context.

There was another well-known example of the I/Os that played out on a national scale: the Mississippi Freedom Democratic Party (MFDP) in the 1960s. In the MFDP's case, a strategy pursued within the civil rights movement changed the terrain on which the whole civil rights struggle itself took place. Their story illustrates how theory develops in real life, how it is needed in order to move ahead but also how it is messy and dialectical.

The I/O Strategy from the MFDP and the civil rights movement

The year 1964 was a crucial one for the civil rights movement and for the nation as a whole. The year before had seen the great March on Washington for Jobs and Freedom, chaired by the president of the Brotherhood of Sleeping Car Porters, A. Phillip Randolph, and which was where Martin Luther King of the Southern Christian Leadership Conference (SCLC) gave his "I Have a Dream" speech. The late John Lewis of the Student Non-Violent Coordinating Committee (SNCC), who would become a congressman from Georgia, had his speech at the March censored for being critical of the Kennedy administration and the FBI. All the scheduled speakers were male, until a protest gave Myrlie Evers (Mrs. Medgar Evers) a slot to honor "Negro Women Freedom Fighters." Subsequent to the March and Kennedy's assassination, the Civil Rights Act of 1964 was passed.

Given that situation, Movement organizations, led by SNCC, decided to up the ante on their previous organizing and pursue a larger, multifaceted strategy in the summer of 1964, both to test and expand the effectiveness of the Act. They called it "Freedom Summer." It included inviting Northern college students—Black, white, and others—to come to the South with SNCC to conduct Freedom Schools, engage in community organizing, and attempt to register Black citizens to vote.

The majority of Black citizens, especially in Mississippi and Alabama but in other Southern states as well, were still barred from electoral participation by poll taxes, literacy tests, state party rules, and threats of retaliation. The Mississippi Democratic Party, which was effectively the only party in the state, had been synonymous with the plantation-owner ruling class of Mississippi ever since the days of the "Redeemers" in the late 1870s, after Radical Reconstruction was overthrown. In the late nineteenth century, state constitutions throughout the South were rewritten to legitimate Jim Crow segregation laws and the disenfranchisement of Black people.

Therefore, movement activists from SNCC working to register Black people to vote in the Democratic Party "white primary" system needed a "safe space" to meet and organize themselves. They would not find that in the regular Mississippi Democratic Party committee meetings or at a convention, like the upcoming 1964 Democratic Party convention where President Lyndon Johnson was the presumed nominee.

Given the exclusion of Black voters from the Mississippi Democratic Party structures, the idea arose to regularize the voting rights movement into a political instrument to both expose the racism of the Mississippi Democratic Party and to provide a vehicle for the movement to express its political demands on a national stage. Thus was born the Mississippi Freedom Democratic Party.

The idea had grown organically from an experience in Lowndes County, Alabama, where a movement organization calling itself the Lowndes County Freedom Organization had arisen, organized largely by SNCC workers and allies. They adopted the Black Panther as their symbol. (This is where the Black Panther Party of Oakland got their name; Huey Newton and Bobby Seale had studied this history.) The MFDP decided to hold an alternative primary election to elect delegates to the Democratic National Convention and to field local candidates for the general election as well. This meant a re-canvassing of existing contacts and many others throughout both the Black community and white sympathizers. This was called the "Freedom Vote." The argument would be that this was the "real" primary, the one where everyone, Black and white, was allowed to vote and elect delegates to the convention. This all took place under conditions of severe terror and repressive, nearly fascist conditions.

On August 24, 1964, the democratically elected delegates from the Freedom Vote arrived at the Democratic National Convention in Atlantic City. They had come at their own expense. They presented themselves to the Credentials Committee with the challenge that they should replace the "regular" Mississippi delegates, because they were the ones who actually represented the Democratic Party voters of Mississippi.

This presented a huge political problem for the leadership of the Democratic Party, all the way up to Lyndon Johnson, who felt that recognizing the MFDP would mean giving up all chances of winning in the South. Barry Goldwater was already running for the Republicans with a program that would allow segregation as a legitimate state-level choice. Therefore, Johnson and the Democratic Party leadership went on full-court press to try to convince the MFDP and its official vice chair, Fannie Lou Hamer, a sharecropper from Sunflower County, Mississippi, who was a leader of SNCC, to agree to a compromise. After much negotiation and pressure from UAW President Walter Reuther, movement lawyer Joseph Rauh, and Senator Hubert Humphrey, one of the most significant congressional allies of the civil rights movement, an offer was made to the MFDP of two at-large seats on the floor as delegates, but retaining the regular MDP delegation. In addition, all the delegates were supposed to pledge support for Johnson.

That compromise was stillborn. Despite incredible pressure from their Northern allies and funders, the MFDP refused the deal. Fannie Lou Hamer said, "I didn't come here for no two damn seats." Her speech to the Credentials Committee was televised. The power of her televised speech was such that, to get her off network TV, Lyndon Johnson personally called up the network heads and scheduled a snap press conference. Famously, he was quoted by his aides writing in their memoirs, "We have to get that illiterate woman off the television." In the middle of Fannie Lou Hamer's speech, which had an audience of tens of millions, perhaps the largest audience that any civil rights leader had reached except possibly Martin Luther King the year before, Johnson effectively ordered the networks to cut her off and come to a press conference at the White House. He had nothing to say, but at least the camera was on him.

Then, ironically, the regular Mississippi Democrats refused to pledge support for the eventual nominee, which everyone expected to be Johnson. Ultimately, the seats on the convention floor reserved for any Mississippi delegation mostly sat empty, since the MFDP had been excluded and the regular delegation refused to sign the pledge. In fact, the Mississippi regulars left the convention and many supported Goldwater.

This was the beginning of the Republican "Southern Strategy" in presidential politics, that was to be exploited by Richard Nixon and Ronald Reagan and was still ongoing under Mitch McConnell and Trump.

When a tactical defeat is a strategic victory

The MFDP neither survived as an organization nor was successful in its challenge to the regulars of the MDP at the convention. Nevertheless, it is a clear example of the power of the I/O strategy because it shifted the entire frame of discussion and action such that in future, no segregated delegation could ever be seated in the Democratic Party. The pressure from this action and the I/OS was such that President Johnson had to support, the following year, the Voting Rights Act of 1965, which resulted, ultimately, in a mass increase in voting among Black people and a large increase in the number of

Black officials throughout the South. A tactical defeat was in fact a strategic victory because the movement made it so.

The "Inside" part was the presence of the self-organized MFDP movement inside the Democratic Party. They did not try to form a separate party and run against Johnson. Instead, through the existence of outside organizing on their own—the "Outside" part—they succeeded in changing the power relations within the Democratic Party.

THE I/OS AND THE CONTINGENT FACULTY MOVEMENT

Theorization of the I/O strategy did not have to start from scratch. Contingent faculty activists like Rich Moser, Gary Zabel, a number of leaders from the CFA including, of course, John Hess, had studied some labor and social movement history as well as having lived it. These were people who knew something about the labor movement and understood how contingent faculty, even those who considered themselves "professionals" and above labor politics, needed to learn that they were part of it.

Within the trade union movement, contingent faculty face a situation comparable to the situation of the MFDP within the Democratic Party. We are officially, but not effectively or appropriately, represented by our overarching union bodies. So the I/O strategy is one for the aggregation, accumulation, organization, and ultimately, the exercise of democratic power by an oppressed group. Then, within the larger body, that power can be leveraged to change the social conditions that led to the oppressed group's second-class status in the first place.

This happened with the contingent faculty movement in 2002 when, at the COCAL V Conference in Montreal, it came time for the discussion of the location of the next COCAL conference, two years later. Activists from Chicago invited the conference to consider their city, where there was a rising level of contingent activity. Top leaders from both NEA and AFT, who were present in Montreal, actively attempted to veto that decision because their affiliates were engaged in competition in adjunct organizing in Chicago at that time. After extended and somewhat nervous discussion, the COCAL leadership decided unanimously to ignore the NEA and AFT leaders, despite the possibility of loss of financial support and possible sabotage of the conference. As it turned out, the 2004 conference was held in Chicago (successfully, and the biggest so far, with substantial international participation), and played a role in (temporarily at least) healing the breach between the two unions. Both AFT and NEA felt it necessary to be represented by high-level officials. They jointly funded a cocktail party and sent staff to observe and find out what was going on. They even allowed themselves to be pictured shaking hands in the middle of a dance floor.

The key to the I/O strategy, in our conception, is that change comes mainly from below, especially change that must be defended democratically for the long term. There is, however, a key difference between the concept of the I/Os in common discourse and our concept of it. We believe that regardless of which part of the strategy is being exercised—the inside part or the outside part—and regardless of which individuals are doing which part of the work at any moment, the decision making should be collectively decided and collectively executed.

The strength of the MFDP and why they exemplify the I/Os of the sort we are supporting is that, when it came down to dealing with a crisis in their *inside strategy*, it was the same people who had run and led the movement *outside*—the Freedom Vote—who sat in that room at the Convention and democratically decided to reject Johnson's offer of two at-large delegates. They did not leave it up to their allies or their lawyer, Joseph Rauh, to decide on their behalf.

Transitory, ephemeral, but also effective

The contingent faculty movement grew out of local and regional movement-building actions, from the 1970s on, for 25 years, and gradually developed a consciousness of what worked and what didn't in advancing the interests of contingent faculty. This was happening at the same time as contingents were becoming the majority of faculty. Generalizing from that experience that created the basis for a national movement, we (with the help of many others—Rich Moser, Gary Zabel, others who have put this idea forward) have theorized the Inside/Outside strategy because we see it as the way that the most successful local and regional struggles have been pursued and the primary strategic way forward for the national movement that we hope we have now demonstrated to exist.

A distinct feature of this contingent faculty movement has been the rise and fall of one "outside" organization after another. This is partly the consequence of the very temporary-ness—contingency—of our employment. It is also a reminder that resources for groups that are essentially oppositional are scarce; the revolution will not be grant funded.[3]

Each of these organizations provided an element of structure and continuity to the movement. Capturing the movement into structures, however, has been like trying to capture one form of a wave that breaks on a rocky shore. The expectation of permanence is probably an illusion. The structure appears, survives for a few years or a few meetings, then declines and disbands. The movement moves on. The secret seems to be that what is permanent is the wave itself. COCAL itself is simply a network that produces conferences; it has no year-round existence between conferences except for an advisory committee, a list-serve and a news aggregator. The budget that supports each conference is whatever is left over from the previous conference, and the network is sustained by volunteer email and Facebook communications.

Another example is the New Faculty Majority (NFM), which was set up in 2009 in order to perform specific lobbying, public outreach, media and assistance to organizing activities on a year-round staffed basis. The discussions around NFM prompted Jack Longmate, a long-time Washington State community college activist, and Frank Cosco, at Vancouver Community College and former union president, to develop a model for bargaining, called "The Program for Change." It explains how through incremental bargaining, contingents could move toward the conditions at Vancouver Community College, where they have nearly eliminated contingency. Of course, that contract was bargained under the more favorable Canadian and British Columbian legal and political conditions. The Program for Change document was disseminated by the NFM and shaped bargaining on the ground in educational institutions around the country.

NFM took as its top agenda item the problem of unemployment benefits for contingent faculty. Toward this end, it undertook a lot of education and some case work where people were working without a union or some other organization to help them. It sparked a coalition among the national unions to get an advisory letter from the Labor Department to clarify for local agencies the actual conditions of contingent faculty with regard to the "absence of reasonable assurance" requirement. This coalition has re-grouped since the Covid-19 pandemic because of the broad relevance of unemployment benefits.

In 2012, NFM succeeded in pulling off a National Summit on Contingent Academic Employment. The summit generated so many reports of contingent experiences that, led by Josh Boldt, it gave birth to the Adjunct Project, a crowd-sourced Google Doc, which was ultimately passed on to the *Chronicle of Higher Education*.

Originally, NFM had been grant-supported. Not only were grants for a fundamentally oppositional organization few and far between, but the constraints of being a 501(c)(3) "politically neutral" organization distorted its mission and ate up personal time and energy from volunteers and staff. NFM decided to stop trying for grants and became a membership organization but as of this writing has not been able to attract enough paying members to support staff for the necessary national coordination.

Motivated by the movement, some adjunct groups arose spontaneously. In 2015, an independent non-profit called PrecariCorps formed to raise emergency funds for contingent faculty. Led by two ex-adjuncts, PrecariCorps lasted into 2017. Also in 2015, an anonymous group of San Jose State Lecturers proposed a National Adjunct Walkout Day, inspired by the Day-Without-Immigrants walkouts.[4] The initiators were anonymous because they were covered by a contract with a no-strike clause. Despite the fear among contingent faculty leaders that the movement was not yet ready for any national action, the idea spread through social media and saw actions on dozens of campuses across the county, ranging from full walkouts to picket lines, guerrilla theater, teach-ins and rallies, often linked with existing unions representing adjuncts or incipient organizing campaigns. Contingent faculty leaders later agreed that they had underestimated the capacity of the movement.

The Coalition on the Future of Higher Education (CFHE), a coming together of major faculty unions across affiliate lines, held a few years of regular meetings and then dissolved when its leadership (largely from the CSU system) stepped back from activity on that level. In addition, there have been various Metro organizations under the COCAL label (Boston, Chicago, New York State, Connecticut and Oregon) that have formed, functioned, and then faded.

In the world of politics, Bernie Sanders' 2016 campaign in the Democratic primaries can be seen as an example of the I/O strategy. It succeeded in altering the terrain, even though it was tactically defeated. In the 2019–20 primary campaign, the strategic victory was on display as the debates all revolved around issues that Bernie had raised in 2016, issues that had been dismissed at that time as extreme or fringe. Fruits of this victory included a joint Task Force with Biden, when he won the nomination and the fact that Biden ran the remainder of his campaign and made his initial appointments substantially to the left of Barack Obama, for whom he served as vice president between 2008 and 2016. The existence of a mass center-left movement, including the Black Lives Matter movement, and the fact that there are now more Sanders supporters elected to seats in the House of Representatives meant that Biden did not have a "honeymoon period" moving on these promises. The Sanders campaign changed the terrain on which the struggle was taking place.

THE RISING LEVEL OF STRUGGLE

However, as if the hands of the clock were moving faster than organization-building can keep up, new directions have emerged as the contingent faculty movement sees its core issues taken up in other forms. The following are examples where fighting in one place changes the fight in other places. What the following have in common is teacher activism.

In San Francisco, the faculty union and students at City College entered into an unprecedented battle, running from 2012 through 2017, against the regional accreditation commission and won; the college celebrated by adopting the label "Free City," meaning getting the city of San Francisco to agree to pay student tuition. If students did not have to pay tuition because they were already eligible for a fee waiver, Free City would provide a stipend for books.[5] This was an idea hatched by the local CFT/AFT 2121 president, Alisa Messer, herself a former contingent activist leader, who was taking a deep breath after a four-year fight. Two years later, "free college" was an item high on national agendas as the Democratic Party candidates vied to beat President Trump.

The SEIU, which thirty years earlier was a union of building custodians and healthcare workers, has decided that organizing adjuncts in private sector colleges is worth a significant investment. The National SEIU made this decision after observing the self-organized efforts by contingents at George Washington University in Washington, DC, who ultimately won a union through cooperation with SEIU Local 500. In their attempt to get a contract and continue their organizing, the Local 500 campaign came to the realization that a Metro strategy was what was needed for them in the DC area. They have proceeded with remarkable success in the years since. The National SEIU took this as a model to scale up nationally and has now been organizing in 25 areas around the country. Under the banner of "Faculty Forward," this campaign has brought 20,000 people into unions, mostly contingents, and mostly from private-sector institutions. These are people who were ignored by the traditional education unions.

A wave of wildcat teacher strikes shook up the traditional teachers unions, sparked by the Chicago teachers in 2012 and continuing into 2019 with strikes in Puerto Rico and statewide strikes in West Virginia, Oklahoma, Colorado, Kentucky and Arizona, states that did not have recognized collective bargaining law for teachers. These have been followed by strikes by unionized teachers in Washington State, Los Angeles and Oakland, California, and a second successful teachers strike by the Chicago Teachers Union. Actual strikes that win demands for the public good have massive public support and make it easier for the public and institutional administrations to believe that more strikes can be coming.

This expansion of the movement has not been unbroken. An independent union at Nassau Community College in New York State went on strike in 2013 and ended up having to settle for virtually what they had been offered originally. But two major higher ed systems, CUNY and the CSU, were able to create a credible strike threat or other credible threat of disruption in 2018 and convince administration, students, and the public, as well as members, that they were capable of following through on that threat. Contingent issues were prominent in both of these struggles. In the 2020 US presidential election campaign, the Bernie Sanders campaign proposed holding public institutions of higher education to a full-time/contingent ratio of 75/25, (a ratio that was in fact passed into law for community colleges in California in 1988 under AB 1725).

a

b

c

COCAL

Contingent Academics Unite!

d

e

1a. Buttons from contingent faculty movement, mainly from the CFA (photo courtesy of Jonathan Karpf).

1b. Buttons and ephemera from separate contingent faculty movement campaigns since 1997 in US, Canada and Mexico (photo courtesy of Joe Berry and Helena Worthen).

1c. Logo from the Coalition of Contingent Academic Labor, used from the late 1990s to the present in various campaigns.

1d. Posters, buttons and news articles from the CFA (photo courtesy of Jonathan Karpf).

1e. Protest against massive class cuts and layoffs at City College of San Francisco, 2021. Former president of AFT 2121 Tim Killikelly and Keally McBride, vice president of the University of San Francisco Faculty Association (SFFA/AFT). The scissors capture the essence of precarity, as contingents are always the first to go (photo courtesy of Helena Worthen).

149

2. Gary Zabel and Harry Brill, early COCAL founders, at UMass Boston in the late 1990s (photo courtesy of Gary Zabel).

a   b   c

3a. John Hess, circa 2005 (photo courtesy of Jonathan Karpf).

3b. The CFA's "Dog and Pony Show" satire on management presentations and programs (photo courtesy of Jonathan Karpf).

3c. CFA San Francisco State University Action (photo courtesy of Jonathan Karpf).

3d. Jonathan Karpf and Alison McKee, Professor of Film Studies in the Department of Radio, Television and Film at San Jose State University and Chapter Vice-Chair for CFA with a "Credible Strike Threat" at San Jose State University. In the background, you can see the famous statue of the athletes, Tommie Smith and John Carlos, SJ State graduates, who raised their fists in a Black Power salute at the 1968 Mexico Olympics.

3e. The CFA at California State Capitol in Sacramento, CA (photo courtesy of Jonathan Karpf).

f

g

h

3f. Mayra Besosa (chair of AAUP Committee on Contingency and leader of the CFA from CSU San Marcos) and John Hess (photo courtesy of Jonathan Karpf).

3g. CFA Lecturers' Council Meeting (photo courtesy of Jonathan Karpf).

3h. CFA leaders from various campuses, delegates to the Edmonton COCAL conference: Chris Cruz Boone, John Boone (husband), Judy Olson, Leslie Bryan, Kathie Zaretsky, Jonathan Karpf, Chris Cox, Meghan O'Donnell, Antonio Gallo at the Edmonton, Canada COCAL (photo courtesy of David Milroy).

4a. COCAL VI march in downtown Chicago to five higher-ed institutions to present report cards on treatment of contingent faculty (photo courtesy of Helena Worthen).

4b. Marchers preparing at Roosevelt U. Joe Berry (with bullhorn) speaking to COCAL VI rally before the march. Linda Turner-Bynoe of CFA is front center with back turned; Rick Packard of CCCLOCC holding sign, center; Tom Suhrbur IEA/NEA organizer, left of Packard (photo courtesy of Helena Worthen).

4c. Joe Berry and Christine Pfeiffer presenting report card at Columbia College, Chicago (photo courtesy of Helena Worthen).

4d. Joe Berry and Frank Brooks presenting report card at Roosevelt University (photo courtesy of Helena Worthen).

4e. CFA delegates to COCAL VI, Chicago. Left to right: John Floyd (Pomona); John Hess, staff; Mayra Besosa, San Marcos; Elena Dorabji, San Jose; Craig Flannery, Los Angeles; Elizabeth Hoffman, Long Beach, Chair of Lecturers' Council, and Linda Turner-Bynoe, Monterey Bay (photo courtesy of Jonathan Karpf).

4f. March stops in front of Harold Washington College. Rick Packard of CCCLOC is center.

a

b

c

d

e

5a. Mexican delegation at University of Alberta, Edmonton. Canada, 2016 COCAL: Sócrates Galicia, Guillermina Velázquez, Horacio Saavedra, Rosalba Flores, Arturo Ramos Almazán, María Teresa Lechuga, Ángel Balderas, Lilia Abarca, José Luis Hidalgo, Rosa Almazán and Ernesto Ortiz (RIP).

5b. Ontario Public Service Employees Union action to gain bargaining rights and contract for contingent faculty and staff in colleges in Ontario (comparable to community colleges in the US).

5c. Action at George Brown College, Toronto, Ontario. "Contract faculty" is a Canadian term for contingent faculty, referring to a limited contract.

5d. Canadian student support action.

5e. Ontario colleges strike action.

153

6a. Plenary session at COCAL X.

6b. Arturo Ramos Alamazan 2004 in Chicago.

6c. Welcome banquet sponsored by AAPAUNAM, and STUNAM. (AAPAUNAM is the *Asociación Autonóma del Personal Académico de la Universidad Nacional Autónoma de México. Syndicato de Insitucion de Universidad Nacional Autónoma de México. STUNAM is the Sindicato de Trabajadores de la UNAM.*)

6d. Open COCAL X International Advisory Committee discussion to evaluate conference and plan future actions.

6e. Sylvain Marois, FNEEQ and Université Laval Québec; Arturo Ramos Alamazán, Maria Teresa Lechuga, David Rives, Oregon Federation of Teachers and COCAL Webmaster; members of the COCAL International Committee (photo courtesy of David Milroy).

6f. Lunch gathering of partial COCAL Advisory Committee in Mexico. Left to right: Jonathan Karpf, Sylvain Marois, David Rives, David Milroy, San Diego; Joe Berry (photo courtesy of David Milroy).

6g. Maria Teresa Lechuga and Mary Ellen Goodwin (photo courtesy of David Milroy).

6h. Marcia Newfield, vice president of Professional Staff Congress, AFT, CUNY and COCAL International Advisory Committee (photo courtesy of Helena Worthen).

154

10
The Contingent Faculty Movement as a Social Movement

> I hope that UPC will break out of the narrow confines of what usually passes for collective bargaining and attack the university on a wide front, aimed at transforming the university into something that services far more people than it does now.
>
> *John Hess, 1981*

Andy Blunden, an Australian Marxist, researcher, and trade unionist, has written a book titled *Hegel for Social Movements* (2019) as a way to make Hegel more accessible to people who are participating in a social movement.[1] The book is written for activists who are likely to have studied Marx as a way of interpreting their own experience, but who have probably been reluctant to invest time in studying Hegel.

Hegel posited that it is through the contradiction or combat of oppositional ideas that new and qualitatively different ideas arise. Marx took Hegel's concept, the dialectic, and applied it to the material world, testing it throughout his study of human social relations, in everything from economics to evolution. Blunden explains how social movements and the simple abstract ideas that animate them are also the product of this dialectical process. In his book, he describes the arc of a social movement:

> To begin with, at point zero, so to speak there is no social movement, there is only a number of people who all occupy some social status or location. But they are not necessarily aware of sharing this status with others. This social status may be being an educated woman, or an African immigrant, or a carrier of the BRACA gene. [p. 61]

Blunden is outlining this process in order for us to appreciate how Hegel's fundamental idea of how change happens through the resolution of rising contradictions lies at the base of contemporary thinking about change, including changes in biological adaptation to the environment, changes (transitions) in human history, changes in collective and individual psychology and of course, changes in social movements.

Or, in our case, someone is hired as a contingent. We might be grad students, research assistants, one-class professors, or half-load or full-load professors off the tenure line, but we would all be hired without job security, or contingent. However, we do not see that as what we have in common. We do not view ourselves as being part of a broad group, but instead assign ourselves individual blame for our status as contingents. At the same time, we question it. We would ask, "Why, if I did all the right things—went to a good university, got an advanced degree, won grants, published research— why am I getting older and older and still don't have a tenure-track job?" The phrase "all the right things" is an indicator of thinking of one's status as being outside of any conflict that might be bigger than the individual level. Of course, doing "all the right things" is not really the ladder that one climbs to grasp the brass ring.

Then something happens which either creates a problem for all the people of this social status or provides an opportunity of some kind. Perhaps only a few people are aware of this fact, or perhaps all of them, but in most cases few of them are aware that the project or opportunity is one shared with others or have an adequate concept of it.

This opportunity started to open up for contingent faculty on a national level in the US in the 1990s, at the time when the MLA Radical Caucus held its first separate gathering for non-tenure-line faculty and when the first COCAL conferences took place. These gatherings were the "opportunity," because they gave people a chance to meet each other, share information, organize, and discuss the idea that the problem was the mounting casualization of the faculty. It was still not widely recognized that this was not just a set of individual failures-to-hire, but a staffing policy that served multiple agendas. The few who were aware took it upon themselves to talk about it publicly and on the adj-l listserve, for example. A few analytic pieces began to appear in book form, in organizational media, and even in the mainstream press. Blunden continues: "Then someone proposes a concept of the situation that they all share, and signifies this concept with a word (probably an existing word imbued with new meaning, or a new word altogether) or some other artifact (such a Gandhi's *charkha*) and publicizes this."[2]

This doesn't happen all at once. In our case, the use of specific words to name our social status tells its own story. Part of the story has to do with higher education administrators trying to find a euphemistic label for this swelling workforce that did not give them a public relations problem that involved contingents gradually replacing traditional full-time professors. In the CSU system, we were *Lecturers*. Lecturers themselves have not challenged that label. In other systems, the name of our group has been a matter of great contention. In the California community college system, for example, we were *part-timers*, because under state law we were unable to work more than 60 percent (now 67 percent) of a full-time load. Some organizers would aggressively use the full legal term, *part-time temporary*, to make a joke out of the fact that one could be temporary for 15 or 20 years. *Adjunct* is the word many of us knew first, especially outside California. This was behind the partial list of these labels in Joe Berry's book, *Reclaiming the Ivory Tower*.

Then in the last few years, along with the words "gig," "precarious," and "zero-hours contract," the word "contingent" came into focus and it became more or less the term of choice. "Contingent" names the common condition that makes good teaching impossible: lack of job security and therefore lack of academic freedom. Thus today we speak about the "contingent faculty movement." The ultimate goal of this movement, though not universally understood by everyone in it, is the elimination of contingency, to replace it with the standard for job security found in other kinds of work: "just cause dismissal". So "contingency" would be an example of naming what Blunden says is "a concept of the situation that they all share." He continues:

> Thanks to this new concept, all the affected or interested people have an understanding of their situation and probably some degree of consciousness of being part of a "group" sharing this situation, and they begin to act in concert in some way. [However,] [n]ew concepts can arise only in institutions where there are definite norms which are capable of manifesting contradictions and stimulating the formation of new concepts. Where "anything goes" there can be no contradictions.

Higher education is certainly an institution with definite norms—for example, the norms of what good curriculum, good teaching, and other good practice look like. The AAUP Statement on Tenure and Academic Freedom is an example of one of those norms. The mission statements produced by every institution of higher ed, even the sleaziest, also boast of "quality," which is another norm.[3] In the daily effort to practice "quality" under degraded working conditions, every contingent has lived under the pressure that shows the sky through the cracks in the contradiction between what is supposed to happen and what does in fact happen. But taking action is not easy:

> To know the problem does not immediately mean to know the solution. Many different solutions will be tried out one after another. The initial concept is thereby modified; misconceptions are dispelled through bitter experience, more and more adequate concepts are formed, until the problem and solution come together in a simple abstract idea which encapsulates the shared problem in such a manner as to be able to transform their situation.

Blunden uses the phrase "bitter experience," and everyone who has been involved in this movement recognizes what that means. We can all think of people whose lives were permanently damaged—or lost—because of the ideas that were tried out but failed, or morphed into something different and left people behind. But once a certain critical mass has been reached, the movement continues. Sometimes it is the very outrageousness of the opposition—for example, the 2012–16 attacks on City College of San Francisco by the ACCJC (Accrediting Commission for Community and Junior Colleges) and its corporate and government allies—that spurs the movement forward.

But—a "simple abstract idea"? It is actually the case that major social movements of the last thirty years have been animated by simple abstract ideas. Examples are the SEIU "Justice for Janitors" campaign and the UPS Teamsters, who in 1997 struck around the slogan "A Part-Time America Won't Work." In 2005, immigrant workers marched to "A Day Without Immigrants" and "We're Workers, Not Criminals." The hotel-room cleaners in UNITE-HERE in the 2010s used the slogan, "One Job Should Be Enough." In Chicago in 2012, it was "The Schools Chicago's Students Deserve." In 2018–19, the teachers in three states struck under "Red for Ed(ucation)." In all of these cases, these simple abstract ideas were useful for people organizing on their own behalf, but also spoke to the public on a recognizably common-good basis, therefore making appeals for solidarity, not charity.

Another simple abstract idea is Medicare For All. The idea of a national healthcare system for the US has morphed from the single-payer (government-payer) program that was supposed to be part of Social Security, to the early days of Medicare in the 1960s, and then various campaigns for Single Payer since the 1980s. Today, Medicare for All, which was part of Bernie Sanders' platform, would transform (actually abolish) the entire health insurance industry. Is today too soon for a comparable "simple abstract idea" for higher education? Is it possible to develop, from the Blue Sky agenda discussed in Chapters 6 and 7, an idea that works both for self-organizing among ourselves and one that engages other workers to act together with us?

Blunden describes stages that are recognizable but our movement has not arrived at these yet:

> All going well, next begins a process of institutionalization. The abstract concept enters general social practice and becomes part of the language. Actions formerly carried out by activists are carried out by institutions— either new institutions (such as a Sex Discrimination Commissioner) or existing institutions are modified (new occupational safety procedures are introduced). Sometimes the activists themselves are appointed to positions in the state apparatus, and even whole activist groups may be co-opted into institutions, such as Aboriginal self-help groups who receive funding for the services they provide.
>
> Through this process the original concept gradually dissolves into everyday discourse and its origin is forgotten. There are no longer any special institutions dedicated to dealing with the issue, but the entire society has been transformed. (p. 62)

The most important example of this historical sequence in the US is the history of the Black Liberation Movement. While the anti-slavery movement dated from the 1600s, it was actions by Black enslaved people themselves during the Civil War, who enacted a general strike, refused to work and then joined the Union Army, that changed the war from being a war to save the union to being a war to abolish life-long racial chattel slavery. That action was animated by the simple abstract idea, "Freedom!" which has been part of the Black movement ever since, through the years of Radical Reconstruction (the attempt to institutionalize revolution in the South) and its defeat, through the years of Jim Crow segregation laws and disenfranchisement, the Second Reconstruction of the 1950s–70s (the Civil Rights movement).

Today it motivates the current movements to create full legal and social equality under the law and in practice.

The contingent faculty movement is at the point where to know the problem is not yet to know the strategy for winning. The real battle—for resources and authority sufficient to turn the course of history around before we burn up the world we live in—will take place on a level above our fight right now. But this shifts us into thinking on the proper scale. The most obvious conclusion to draw is that the transitions we need to bring about are not going to be won by individual local unions at the bargaining table. They go far beyond that. A comparison between the Blue Sky conditions envisioned in Chapters 6 and 7, "What we fight for" versus what is captured in the CFA contract should show how incommensurable, yet how closely linked, they are.

However, at the level of our rising national movement, we can point to examples of how our efforts have led to changes that have become institutionalized in some respects. We have recognition by major unions, action plans, and strategic initiatives flowing from them; legal changes that have been pressed and won in legislatures, courts and other government bodies, and the spread of collective bargaining at a time when collective bargaining is itself under attack. The movement is also much more conscious of its place in the broader struggle within the political economy of higher education.

We are entering this dialogue at a moment when a lot is at risk, but a lot is possible. The rest of this book is intended to be a guide to navigating that moment.

PART V

SEVEN TROUBLESOME QUESTIONS

11
What Gets People Moving?

Starting with this chapter, we try to open up key questions that come up in the course of organizing and representing contingents. These are neither global philosophical questions like "Why organize?" nor practical contract bargaining or procedural questions.

Instead, these are troublesome, self-reflective questions that rank-and-file organizers ask themselves about their own work. Typically, they occur to one individual and then that person finds that other people are asking the same question. That realization starts a discussion. Often, the rank-and-filers bring these questions to a professional staff organizer. However, that outsider's clear professional answer—or worse yet, a quick answer to any of these questions—is not always the best answer. A clear answer, no matter how "correct," can stifle discussion, even if it is not meant to. This point was one of the central themes of Myles Horton's career as a working-class educator.[1] These questions are best answered in practice, in an ongoing way, by working faculty talking with each other in the midst of organizing. They are likely to be the kind of questions that are best discussed in a safe space among people who share a problem equally.

We will try to avoid answering these questions. Our intention is to make them troublesome, not to answer them. Our assumption, to paraphrase Parker and Gruelle's book *Democracy Is Power*, is that they can be tools for building strong relationships within an emerging union.[2] Working faculty need to be able to think and talk through these questions if they are going to create a democratic, participatory non-staff-dominated union. This is the internal corollary of the external "fighting for the common good." Both have broad democracy as their common denominator.

These are also all questions that really did arise in the course of the very rich, relatively successful forty-year struggle of the Lecturers in the CSUs. In addition, we are going to draw on other organizing efforts that have taken place over the last thirty years.

First, what gets people started and what keeps them moving? Getting someone started may meet resistance. Often, the first outreach is an invitation to a meeting. But an invitation to a union meeting may be met with this remark: "I don't get involved in politics." This person does not mean Democrat-Republican politics or electoral politics generally. They mean that they want to remain unknown to the powers that run their workplace except as an academic, a teacher, researcher, committee participant, etc. They want to postpone as long as possible being marked by colleagues or bosses as a contingent, especially as a contingent who is unhappy with their working conditions and may make trouble. So "I don't get involved in politics" means "I want to avoid conflict," and the assumption is that coming to a union meeting of some sort will inevitably draw them into conflict and make them visible for that reason to management. We've told the story of how John Hess, when he first came to San Francisco State, tried to keep his head down as long as possible. "Come, teach your class and leave! Because if you don't do that, it would inevitably piss somebody off."

However, sooner or later, the very conditions of their work, the impossible choices that contingents are forced to make between doing it right and doing it the way they are pressured to do it, will turn this person toward the union or the need for one. Eventually, conflict will find them and they will be open to an invitation. In the meantime, the union has to keep the invitation active and repeated. Elizabeth Hoffman had been drawn into the UPC struggle and then resisted joining the CFA until the staffer found her in her office and talked her into signing, and then ten seconds later got her to run for office. Within a few weeks, she was leading a meeting. A well-known AFT research study of many years ago found that it took an average of seven communications ("touched seven times") to bring someone into the union who had not been predisposed to it before.

THE WAY A MOVEMENT SWEEPS PEOPLE IN

Hopefully, the meeting that a newcomer will come to is neither just purely for business, or just "pissing and moaning," as John Hess put it. The meeting has to be the right combination of sharing personal experiences, planning for collective action, and evaluating past collective action.[3] The newcomer has a role to play in this. When successful organizing gets going, a movement develops, gathers energy, and broadens over an extended period of time.

It brings in new people who have different experiences. They are different from each other, but involved in the same thing. The movement is a culture, growing and changing. It has momentum.

People say that that when the number of phone calls coming in, or from people whose names you don't know, exceeds the number going out, you know you've got a movement. It's not just quantity; it's also that the demands from the people at the bottom become, over time, more expansive than the demands of the leadership. This was famously the case in the Montgomery Bus Boycott of 1955. The struggle started out to bring partial relief from the degradation of being shunted to the back of the bus even when there were empty seats in the front. It became a demand for total desegregation and the hiring of Black bus drivers.

But individuals do give up some of their autonomy in order to share in building this movement culture even as it changes. How does an organizer encourage, and support this at the same time as channeling it?

HOPE AND FEAR

Imagine the consciousness of our contingent colleagues as a set of scales: on one side of the scale is fear of consequences, negative impacts if one takes action, and fatalism—the sense that nothing can be done to change anything. On the other side of the scales are courage and anger at the injustice of our situation, and hope and vision that a change, a new or altered system, is possible. The scales rise and fall as people's balance of hope and fear changes over time. So the organizer is always trying to move one grain of sand from the fear-and-fatalism side of the scales to the hope-and-courage side.

Luckily, this is not a purely individual matter. Small victories get reinforced and minor setbacks become shared lessons. Moving that grain of sand from one side to the other creates motion. Newton's First Law of Motion, which is about inertia, applies to social relationships, too; once people get started, they tend to keep going. To extend the metaphor, we can add the concepts of momentum and acceleration. People's relationships, the web of support that is created by the actions of being together in a continuing struggle, grow stronger. Each motion makes it easier to recruit the next person, to move the next grain of sand to the other scale. The shift of the balance begins to accelerate. One plus one equals more than two, and two plus two equals more than four.

To be sustained and successful, a movement must have a central idea. This central idea has to be made explicit and conscious, so that people can imagine it. They have to be able to imagine, for example, a world in which, as C.L.R. James put it, paraphrasing Lenin, "Every cook can govern." In the case of contingents, the idea is a workplace in which they are treated with appropriate respect and equality, but even a basic idea about what this looks like in practice has to be spoken out loud and explained. What does real respect for the work of faculty in higher education look like? What are its essential features?

Take the example of justice. We are not born with the concept of justice; the conscious, explicit desire for justice is something that has evolved through culture. But culture is not individual. It is a collective, historical phenomenon. No one has a "culture" all by themselves.

Take another example: slavery. At what point did people start to say that freedom was the fundamental right of all human beings? It was at that point that the idea could become embodied in a movement. Slavery goes back to early human history. Slaves certainly always wanted to be free themselves, but although there were slave and peasant rebellions—Spartacus in ancient Rome, Bacon's Rebellion in Virginia in the 1600s—there was no anti-slavery movement. There was no general revolt against the idea of slavery or against being part of a culture that allowed slavery. Freedom was about individuals. In fact, there were many cases in history all the way back to the ancients where slaves became free and then became slaveowners themselves. This happened even in the very early British colonies, where some Black former slaves became slaveowners.

The idea that freedom was a human right and that slavery was an evil for everyone in a slave society broke into modern history with the successful 1791 slave revolt in Haiti, led by Toussaint L'Ouverture. Inspired in part by the stated ideals of equality of the American Revolution (1776) and the French Revolution (1789), this revolt against the French colonial rulers abolished all slavery in that small country. Before this, there had been rebellions undertaken by slaves to free themselves, but no other recorded cases where it was a genuine revolution to change the social relations of a whole society. Only after the revolution in Haiti did we see rising movements in the US and Europe against the very idea of slavery, envisioning a more just society in which *no one* could be a slave. In the US, the anti-slavery movement began to take shape at that time.[4]

Therefore the central idea of a movement has to be explicit and conscious. But it also has to be different. Moses may have said, "Let my people go," and the Hebrew slaves walked across the Red Sea to freedom, but that was not the same as "End slavery!" How does this apply to higher education faculty? If the practical idea around which current organizing is taking place is the end of contingency—job security and academic freedom for all higher education faculty—is that enough? Is it the "simple abstract idea" that Andy Blunden looks for? Contingency is an evil that permeates all types of workplaces and therefore can be fought on many fronts. This creates the opportunity for us to be part of a movement broader than the structures of higher education.

A NEW SOCIETY IN THE SHELL OF THE OLD

The relationships built in the course of the struggle can prefigure the kind of change we want and the kind of society we want to live in. In the old days of the Industrial Workers of the World (IWW), it was said that "The Union is the New Society in the Shell of the Old"—that is, today's organization in struggle prefigures the new. Now we are talking about union democracy. For many people, the experience of union democracy is the first time they have ever encountered democracy in practice. It can be an eye-opener. When activists carry out their work efficiently and effectively, with a sense of clear purpose, the stark contrast between the culture of the union and the hierarchical culture of the institutional dictatorship in which most people work is an education in itself.[5]

Many kinds of collective decision making take place in unions: meetings, sharing of resources, the relatively flat hierarchy, a respect for information no matter who provides it, and, above all, relative equality. These are in themselves educational, especially when these are done well, promptly, and with accountability. "We can do better than they can," becomes the sense of the movement; we can run our institutions better than our bosses can. This pride and self-confidence can infuse both the organizing effort and, once the union has been established, the whole relationship between the union and management.[6]

People become committed to the positive, affirmative social relations within the union. This contrasts with the hierarchy of the institution where they work, where social relationships are starkly unequal and competitive. In the union, they are all members, with leaders elected from among the members and supported by staff. We will bring this issue up again when we talk about how unions diverge from this ideal and what we can do about it. This is important because in the heat of a struggle, what people rely on are immediate relationships. Soldiers may enlist for an idea, but in the field they do not fight mainly because of some abstract idea. They act with courage and bravery because of their strong personal relationship with the person fighting next to them.

This is, of course, an idealized picture. However, even if it only exists briefly—even if it is only a phase through which a union passes on its way to being an ordinary business union—those who have experienced it will be thinking of this phase when, years later, they look back and say how they loved their union. "It turns out they had a community and they didn't know it," said John Hess, of people he brought into the union.

Who says they don't care?

Many contingent faculty teach at night, and some teach only at night. They work in buildings where the lights are off in many rooms and where the corridors are empty when class is over. Parking lots are dark and lonely. Buses have finished their last runs and janitors are waiting to empty the wastebaskets and lock the doors. Some of these people have day jobs that are in different fields besides teaching. Some are managers in these other fields and may be even anti-union. Most of these people have no relationships with anyone at the school except the person who hired them and their students. The union is the only medium that can break these people out of isolation and provide equal, supportive relationships in the context of their job. But until they are reached, they not only do not know the union—the union does not know them.

People who teach at night or in other isolated situations are a challenge for organizers. John Hess was volunteering to organize Lecturers for UPC in the run-up to the collective bargaining election in 1980. John Hess's union president at the UPC said, "Don't bother talking to the night part-timers. They don't care." This was a mistake, part of a general attitude that would contribute to costing the UPC the representation election, and it was not true. Many people, especially other Lecturers, reached out to the people who taught at night anyway and found that everyone cared. Even those who were managers in their day jobs were contingents when they were teaching, and they experienced management's bite of disrespect for their work and for their students. But when the union said, "They don't care," that was really the union saying, "We don't care about them."

The union, in attempting to create its own culture of a different world, has to care about all potential members. In reality, all the Lecturers demonstrated that they did care and they made that union and its successor, the CFA, care about them.

A similar example happened at AFT 2121 at City College of San Francisco (CCSF) at in the early 1980s, with Joe Berry's own local president, Steven Levinson. The context was internal organizing prior to an agency fee election. Levinson was reluctantly induced to spending an evening talking with the part-timers who taught at night. He came to campus one evening and met with them in the mailroom as they came in to pick up their mail. Joe introduced him. Every one of them who was not already a member joined the union. Levinson, who had also believed that the night workers didn't care, changed his mind that night.

Everyone needs to be individually organized

Everyone agrees that people have to be approached individually to be brought in. This is true even of people who were political radicals or activists in support of other people previously—like John Hess. This may seem overwhelmingly time-consuming. Isn't there a shortcut? No. Why not? Because it's all about relationships.

We have found in labor education classes that a very effective way to bring home this idea in a class on organizing is to begin by having people sit in a circle and describe how they themselves were first brought into union activity. Overwhelmingly, they say they were asked, and they can name the person who did it. Getting people to think in terms of how they were first brought in, and how that felt, which may have been years before, is an important first step in getting people to think like an organizer.

An activist on your own behalf: Put down your bucket where you are

It is fundamentally different to be an activist and organizer on your own behalf, especially on your own job where your own paycheck is at risk. A staffer who has a good job with a union is not in danger when they organize someone else. They are in safety, looking on while a worker is exposed to the gaze of the employer. This is a point that John Hess made repeatedly. Getting active in the union was not automatic, even for him. The first time he was called to go to a meeting his first reaction was fear that he would "piss somebody off." Organizing as a Lecturer on his own campus did not come automatically even though he was already a socialist and general activist.

There are at least two factors to consider here. One is simply the fear that any worker feels when being tempted or impelled to step forward and speak out, or organize on their own behalf as workers, individually or collectively. This is much scarier than the more militant actions that might be taken by an activist or a socialist on behalf of people in another time, place, or situation.

Our workforce may be uniquely trained to avoid conflict in our professional lives in general. We trained to resolve things in logical discourse. We do not easily perceive ourselves as exploited labor. We are operating, after all, at the top of the working class in terms of education and cultural appropriation, doing social reproduction close to if not at the very point of production of cultural hegemony. At the same time, we are far from the top in economic rewards and working conditions. Leading our colleagues to recognize these facts and act upon them without liberal guilt is an especially important task for organizers in this sector. Just as we need to challenge our own role as workers, we also need to challenge the role in which we are cast, as reproducers of the cultural hegemony that the higher education industry manufactures for the purpose of maintaining class power relations. Teaching one's students about organizing, labor history, left politics, or even socialism is not the same as organizing among ones own colleagues.

Unfortunately, we have seen many counterexamples where political radicals and activists in other venues refuse to be active on their own jobs, or even join the union at all. What explains this? This is a troublesome question.

Positive reinforcement and respect for the actual work of contingents

A steady supply of positive reinforcement from someone who is a leader, officially or unofficially, is what moves the grains of sand from one side of the scale to the other, but positive reinforcement is also powerful when it comes from someone who has no other authority than their personal relationship with you. Helena Worthen got a lesson about positive reinforcement in a particularly focused way by Billy Henderson, a past president of her local, AFT Local 1603. This was in 1989, in the middle of the AIDS epidemic. Henderson was at this point dying of AIDS, although it was a secret. One night, Helena, who did not know about it, found out when she was asked to pick him up from the hospital where he was undergoing treatment. He invited her into his home and made her a cup of tea. As a thank-you, he gave her a piece of advice. "You can do anything using positive reinforcement," he said. She was dubious—anything? Anything at all? Really? Well, probably yes. She has been testing this claim ever since.

Positive reinforcement is not just personal. A celebration of the work of the people being organized can be experienced as positive reinforcement for the whole group. This builds an organizational culture that encourages more activity and initiative and pays respect where respect is scarce. This is especially important to contingent faculty, who get no respect from anyone but our students.

Events that are not directly about organizing or the union, but are about the subject matter and disciplines that faculty have chosen because they love them, provide positive reinforcement on the basis of the very work that receives little respect in their parent institutions. A conference or festival in which people share the content of their teaching with each other reflects the kind of opportunities that full-time tenure-line faculty access when going to national or international conferences, but it has a local organizing function as well. For example, during a campaign to organize the faculty at the Art Institute in San Francisco, where all the faculty were contingent, Jessica Lawless, an artist and former contingent herself who was on the staff of the SEIU, supported the faculty in creating their own art festival. Since they were all artists, the opportunity to have a collective show released a tremendous amount of creative energy. The festival was successful and was expanded to an organizing campaign in Los Angeles.[7]

12
Who is the Enemy? Who are Our Allies?

We talk about fights, tactics, struggles, wins and losses. We use language that suggests war. At the same time, we keep working, doing our jobs as well as possible in the same institutions where we are doing our fighting. Teaching is a labor of love. So who is the enemy? In each case, the definition emerges in practice. We must construct that definition in order to assess our own strengths and weaknesses as well as theirs.

OUR IMMEDIATE SUPERVISORS

When looking for the enemy, we are likely to start with the people who are directly in charge of our working conditions on a daily basis. These are department chairs, program directors, coordinators, assistant deans, full-time tenure-line faculty, and sometimes even other contingents who have been delegated to implement the system. They have a great deal of power over our immediate work lives. On the one hand, they hire us, evaluate us, reward us with assignments, and can make life fairly easy for us. But they are also the people who can deny us a library card, ignore the contract, discriminate in giving out class assignments, "de-schedule" us, ignore us in making curriculum decisions, cancel classes before enrollment is complete, read but fail to investigate our student complaints, tell us where we can park, make decisions that affect us without asking for our input, and disrespect us on a daily basis. This is a close, complex relationship, charged with difficult emotions. But it is a mistake to assume they are necessarily the enemy, or the main enemy. Their power has a higher source.

THE NEXT LEVEL UP

Our immediate bosses are bound by constraints they only partly control. They are themselves employees who get their power from the structure of higher education institutions. We need to trace the power train up, through deans, vice presidents, provosts, boards of trustees, donors, legislatures and big capital itself. All the way up this power train, our precarity serves some purpose. In different ways, someone at each level benefits from it. Eventually, we end up with a fairly small group of people who govern not only our own institutions but higher education in general as a part of the country's political economy. We rarely get to meet these other people.[1]

As Alisa Messer said in 2013, when she was president of AFT Local 2121 at City College of San Francisco (CCSF) during an existential struggle for the survival of the college, "We have a hard time putting a face on the enemy." In order to see clearly who these people are and what their interests are, we have to draw the class line. Then, when we raise the level of struggle to come face to face with this higher level, we find that some people at the lower levels can be our tactical allies. An obvious example is when the fight is about adequate state funding for public education, as happened during the wave of teacher strikes in 2018–19. In many states, school administrators and teacher unions found themselves on the same side. In a healthy union, this change in our thinking about who is our enemy, when and where, can be openly discussed, but it takes organizing to make this come about.

RAISING THE LEVEL OF STRUGGLE: THE EXAMPLE OF GRIEVANCES

A lot of the daily work of a union revolves around things like enforcing the contract, participating in union committees, making presentations, handling grievances, etc. This can feel like a hamster-on-the-treadmill trap unless we think of it as a way to lift our attention higher, past the immediate manager, to the source of the power that is exercised over us. What to do with a grievance is a question that can open up this discussion. The limits of the lower-level administrators' power can be revealed when we raise the level of struggle.

For example, we can resolve an individual grievance and help make whole one worker, but some injuries really affect all of us. We can file a collective grievance about this on behalf of the union. Collective grievances are also public grievances. They become news both for the membership and for the enemy at all levels. This balances the limited powers of low-level administrators with the equal and rising power of the union. The union may lay out a path of least resistance for whatever administrator is immediately responsible, but if that person's powers are limited and they can't walk down that path, the whole issue can be pushed up another level. This can reveal what interests are at stake for those at the level above. It requires some kind of leader, however, to perceive this and point it out.

The Australian Mike Newman, a leading adult and union educator, titled one of his books *Defining the Enemy*, which underlines the importance of making "Who is the enemy?" a key strategic question. He says that the purpose of adult education is to define the enemy concretely, down to giving an actual name and address. We try to define and redefine the enemy continually and publicly, especially to our colleagues and fellow workers, because that's an essential part of building a movement that can change power relations.

WHERE THERE ARE ENEMIES, THERE ARE ALLIES

When we try to define our allies and enemies, the question is not "Is this person a friend, or even a good person?" The question is "What is their interest?" Being clear about this is critically important to us as purveyors of a public good (higher education), especially those of us employed in the public sector. This makes our need for allies different from that of some other workers—for example, those who work in the private sector, such as manufacturing, where production is primarily for direct profit. But just because we provide a public good does not mean that other workers see their interests aligned with ours. Alliances do not come automatically. They have to be built.[2]

Building a broad base for public support in our struggles is very important. You would also expect it to be possible, since "our working conditions are our students learning conditions," and our students themselves are a mass base. However, we don't have power like that of workers who produce refrigerators, hamburgers, Internet communications, transportation and other commodities. These workers can directly impact the profits of capitalists by withdrawing their labor; we cannot. The main aspect of our work—teaching—does not directly produce private profits except in for-profit colleges. On the other hand, should we come to the point of withdrawing our labor, as in the 2017–19 wave of teacher strikes, our mass base and their families experience a major public inconvenience.[3] The potential backlash from this can best be managed if alliances are well established in advance. Our best appeal to them is as fellow workers in an essential industry doing the necessary labor of social reproduction.

In the months and years before their 2012 strike, the Chicago teachers of AFT Local 1, the CTU, researched and wrote a report published as a pamphlet called "The Schools Chicago's Students Deserve."[4] It made the argument for how good schools for Chicago's 400,000 students were not only needed but possible, and also clearly identified what obstacles would have to be removed: the school district administration, and Mayor Rahm Emanuel and his appointed school board. One effect of this document was to bring parents and students into alliance with the striking teachers. The nine-day strike was a citywide celebration of support for the teachers. People said, "They turned Chicago red," referring to the red T-shirts and "Red for Ed(ucation)." After winning substantial improvements in the 2012 contract they struck again in 2019 and won another partial, but substantial, victory over the new mayor, Lori Lightfoot, who had postured as a friend of education.

Dividing, isolating, conquering

One way to identify the enemy is to ask who is attempting to divide us and how. Is some institutional practice encouraging competition among faculty, or a sort-and-discard approach to grading students? It may seem as if tenure-line faculty are the enemy if they are assigned to evaluate or supervise us. Or it may seem as if students are the enemy if student evaluations and complaints are used as the sole criteria for evaluation. But who designed that policy? Neither tenure-line faculty nor students run the system. They are instruments given these assignments, so we have to look beyond them to see where the assignments are coming from and who has the actual power.

The distortions of human development under capitalism are things we encounter every day in our work.[5] We see this in the sharpest form in the for-profit part of the higher education industry. We have to look past the distortion to find the original, human connection, which is where we will find our allies.

Campus labor coalitions

When we scan the categories of employee groups at institutions of higher education to find out who is engaged in struggles like ours, we can quickly see some potential allies. In Herbert and Apkarian's 2019 article summarizing work stoppages in higher education, the most common group to carry out work stoppages is non-academic workers, who carried out 21 out of a total 43.[6] Between contingent and tenure-line bargaining units, there were 14 work stoppages (contingent faculty units six, combined units eight) and graduate students, who have many fewer units total in the country, carried out seven.

Graduate employees—research assistants, post-docs, teaching assistants who often carry large classes and function as faculty of record—have been organizing actively over the last thirty years. They have recently been joined by undergraduate student workers. Interestingly, this experience has led many students to decide they would rather be union organizers than professors. Some have become both. In places where they have organized, they have often played the role of sparking a higher level of activity among other unions and new organization among other workers, and generally providing the youthful energy that has made the campus labor movement one of the most encouraging spaces in the whole labor movement. Their ongoing legal struggle to gain employee status in the private sector and in many states is one of the longest running legal issue battles that the labor movement confronts. For contingent faculty in particular, relations with graduate employees have not been untroubled, since many grad employees, especially at elite universities, still see themselves as likely on the path to tenure-line jobs and are easily led to see us as the "failed ones."

Workers on any campus include custodians, drivers, machinists, food service workers, clerical workers, all the building trades, landscapers, non-management administrators, and just about every other kind of work necessary to a society—and this is in addition to the many kinds of academic workers. Many of them are unionized. Some are simultaneously students taking classes while working. A campus labor coalition can bring the strength of these different groups together for mutual support. These are long-term projects that pay off in terms of years and even decades. "Mutual" means that faculty cannot expect the support of the custodians unless the custodians can expect the support of the faculty in their own struggles. In order for the labor coalition to develop power, the faculty cannot expect to act like teachers or wise older siblings in this context. Faculty's readiness to learn the complex and significant lessons from the work situation of fellow workers can make or break a campus coalition. Examples of faculty's failure to understand their proper role in these alliances are unfortunately many and historic. A labor education or labor studies program can provide some support and direction, since it will have roots in both labor and academia.

Tactical alliances: "We walk with them as far as we can"

A good point about allies was made by a leader of the Iraqi Oil Workers union. After the US invasion of Iraq in 2003, a group of US labor organizations and individual rank-and-file activists made an effort to speak in labor's voice against the war and build solidarity with labor and working-class forces in Iraq. As part of their organizing effort, US Labor Against the War (USLAW) set up an exchange. Iraqi trade unionists came to the US twice and a delegation of US trade unionists went to Iraq. One of their events took place in Chicago at the United Electrical Workers Hall (UE) on Ashland Avenue. The Iraqis described how they opposed the takeover and subsequent denationalization of the Iraqi oil fields by the US occupation forces. To do this, they worked with their bosses in the Iraqi oil industry. An American asked how the union could work with their bosses. The Iraqi oil union leader said, "We walk with them as far as we can."

This is a very clear example, under extreme, even life-and-death, conditions, of the distinction between tactical allies and opponents and strategic allies and opponents. The oil industry management were tactical allies for Iraqi workers against US imperialism and privatization, but US workers and their unions can be strategic allies. For us as contingent faculty, the same might apply to our tactical alliances with university and college management, especially in the public sector, when we fight for adequate funding from government while at the same time recognizing that ultimately, our interests strategically diverge from management as a group. This is not to say that individuals do not sometimes pursue a different course, but their survival as managers, if they do that, is usually limited.

What side are you on? Class consciousness and class analysis

If you know you are making a tactical alliance, you must clearly envision and plan for the moment when the path along which you walk together separates and you go in a new direction. As leaders, your plan for this change must be made clear in advance to the rank and file. Failing to do this puts the leadership at risk of posing for photographs smiling with management at one point and then being quoted as adversarial the next day—a shift that causes confusion and cynicism among the rank and file if they are not prepared for it and were not included in the strategic decision making that led to it.

There is a famous poster from Pressgang Publishers (Toronto, 1978) that hangs on many walls in the homes of people who worked in the civil rights, anti-war, and other movements. A working-class woman of indeterminate race is leaning on a fence that probably separates two backyards in a working-class neighborhood. She's frowning but calm; she is thinking. The words printed below the image are "Class consciousness is knowing which side of the fence you are on. Class analysis is figuring out who is there with you."

Another way to talk about class is to compare two ways of picturing it. One is "the class in itself" and the other is the "class for itself." The idea of a "class *in* itself" defines a class based on its structural position within the economy: who is exploited, who depends on wages to live? The idea of a "class *for* itself" asks us to see ourselves as workers and find alliances for ourselves among groups of other workers who see themselves likewise. The first way refers to material position in the economy; the second is how people think of themselves, a matter of consciousness.

Students as allies

The first circle of possible allies is our students, the people we are alone with every day in the classroom, sharing the same conditions and limitations that we experience. They are also often the closest relationships that we develop, since many of us as contingents have to make a special effort to develop relationships outside our classroom.

There are two key factors that can point us in the direction of effective alliances with our students. We need to come out of the closet to our students, who otherwise have no way of knowing that their teacher is a per-class contingent and not a professor with a capital "P" with tenure and a very different life. In addition, we need to recognize that we and most of our students are part of the working class and are fundamentally on the same side—our side—in the world. Once these two conditions are met, experience has shown that students will rally in large numbers in our support, often with creative ways and impactful results.

Immediately after the Revolution in the CFA, it was the faculty of color in the CSUs who had a particularly close and sympathetic relationship with the students of color on the campuses. They proposed to the CFA that the union strategically prioritize organizing this growing group of students and students in general. Both John Hess and Susan Meisenhelder, who was CFA president at the time, saw these alliances as crucial in the development of the CFA and its leadership after the Revolution of 1999. Organizations of faculty of color within the CFA were themselves engaging in an Inside/ Outside strategy arising from their similar positions as second-class citizens in academia.

The CFA budgeted money to supply these students with a small stipend so that they would have paid time to do this important work. Paying students was particularly important in a mostly working-class system like the CSUs, where the majority of students were not only borrowing money to go to school but also working to pay school and living expenses. On every single campus, students began to organize students, developing coalitions with existing student groups in support of the common struggle to "save the CSUs."

In order to adopt this ultimately very successful strategy, the CFA had to discuss it internally and overcome objections from some faculty, who were concerned that the union paying students to organize other students was an abuse of faculty authority over students. This was a complicated matter. The objection correctly called for serious discussion, because faculty members and teachers in general, just like parents and other authority figures, can easily abuse, and be seen as abusing, their power over students and other younger or otherwise less powerful people.

In this case, the CFA decided that the specific tactics and issues that the students organized around would be decided by the students themselves. The faculty would only give advice when they asked for it. The result of the discussion was support for the proposal, since the primary organizers of students were not going to be faculty but other students. There were also obvious parallels to paying Lecturers to do work for the union to free them from having to work a second or third job, a tactic both the CFA and other unions have adopted to their advantage.

The discussion and acceptance of this proposal helped, over the years, to solidify support for the union among faculty of color, whose support could not be just presumed. It also made the union stronger and enhanced the position of the faculty of color within the union. This was a clear example of the Inside/Outside strategy working successfully.

Are tenure-line faculty allies?

It is easy and tempting to see tenure-line faculty as a group as the enemy, but it's a mistake. Certainly, the most challenging alliances for contingents may be those we make with tenure-line faculty. The contrast between the conditions under which they do their work and the conditions under which we do our largely identical work is stark. This includes tenure-line faculty supervising us, evaluating us, and choosing to take classes (as in summer sessions and through overload) which we would otherwise teach. Contingents have to manage this relationship individually and collectively every day. As a relationship that is both intimate and power-based, it can be compared to a dysfunctional family relationship, except that it cannot be fixed by therapy. This relationship is very close to home, as equally close to home as that with students.

Our key consideration in dealing with tenure-line faculty should be that the strategy for contingents should not be structured around an anticipated or even an actual reaction from tenure-line faculty. They will have their own ideas about what contingents should or can do, and they will expect to be listened to. Listening is one thing, however: giving over power is another. In designing strategy, contingents have to focus on our own capacities and strengths, not on how our actions will be interpreted or reacted to by tenure-line faculty. When we organize ourselves and make our own decisions, we can have relationships of solidarity rather than be grateful recipients of fleeting condescension.

Effective strategies for contingents will likely violate the expectations of tenure-line faculty. A pivotal point for John Hess was his epiphany after realizing that the tenure-line faculty organizations were coming to the Lecturers' Council meetings in order to learn. He heard himself say to himself: "They have nothing to teach us." There was no turning back after that.

13
What is "Professionalism" for Us?

"I don't need a union. I'm a professional!" We have heard all kinds of people say this. It is a vision of professionalism that limits collective action and is ultimately anti-union. It reflects the attitude that a practitioner of some profession (or craft or art), once educated and professionalized, often with some sort of legal or extra-legal certification, becomes an independent professional whose class position should be based on their individual merit or expertise. This is despite the fact that most professionals, even in the traditional professions such as medicine or law, are now effectively employed as wage and salary workers.

This is overwhelmingly the case for those of us in higher education. The old ideas die hard. Granted, they are based in a historic reality: in the days when the majority of higher-ed teachers were tenure-line professors, many also had, in addition to the property rights to their job as faculty, professional knowledge that they could market as consultants, authors, or public intellectuals. This allowed them to function as independent professionals in the sense of running a small business based on their expertise. They were simultaneously faculty, earning a wage, and independent professionals earning fees, which placed them in the "contradictory class position" described by Erik Olin Wright, a position that could make it hard to identify with the historically working-class idea of unions.[1]

Both the National Education Association (NEA) and the American Association of University Professors (AAUP), of course, resisted calling themselves "unions" and used the word "association" well into the 1990s, for exactly this reason. The historic NEA was dominated by education-school professors, upper administrators of schools and colleges, and their allies in business and government. The AAUP, as a limited guild of prominent professors, also reflected this perspective. Originally, one could enter and become a member of the AAUP only on the individual nomination of someone who was already a member. The Congress of Faculty Associations (later re-named the California Faculty Association) in the CSUs was the child of this heritage and won its election on the basis of being the "non-union" collective bargaining agent.

But these old ideas also persist because the cultural prestige of independent professional middle-class status is part of what has historically motivated students to aspire to these positions and pursue years of education, often at great cost to themselves and their families. To be a professional whose job places one close to the top of the source of cultural hegemony can be intoxicatingly satisfying. To give up that upwardly mobile aspiration and acknowledge one's present and most likely future status as part of the waged working class requires overcoming the cognitive dissonance that only a general change in consciousness can accomplish.

How that change in consciousness takes place is the real concern of this chapter. For example, it is not just a matter of how contingent faculty privately view themselves. Many are reluctant to be public about their contingent status with their students, friends, family (who may have co-signed student loans), or even colleagues. For some, it takes the start of a public campaign to give them the confidence to "come out of the closet" of contingency to be public about their status.[2]

A different aspect of the same basically elitist attitude was on display when John Hess heard the shout, "Throw the Lecturers out of the union!" This attitude exhibits the "guild" consciousness of restrictive entry into the group as being the key marker of professional status. Therefore, keeping other people out of the group, at least officially, if not effectively in practice, becomes an ideological and psychological necessity for those who have achieved full guild membership, or in this case, professorial perks and status. In fact, there are unions in higher education that have called themselves a "guild" in their titles—for example, Los Angeles, San Diego, and Glendale community college district unions—long before contingents became a substantial political force within them.[3]

Versions of guild consciousness often arise with regard to contingent participation in faculty senates and other institutions of supposed shared governance between faculty and management. It is the same attitude that led many other unions in other sectors (railroads, skilled trades, etc.) to exclude Black workers, women, Asians, Latinx, and others. The fear is that if they are recognized as full members of the craft and guild, then the entire trade is devalued, since they are inherently less valuable people. Managers in higher education foresaw the value of guild consciousness and welcomed it because it justified their devaluation of faculty jobs as the hiring pool diversified, made the budget go further, and otherwise facilitated solving their four problems described in Chapter 5.

These linked visions of professionalism exist on the basis of drawing the line between professionals and other working people. This line can be drawn even when the professionals in question have been stripped of their economic independence and rely on their employer for access to both their necessary means of production (that is, classrooms, and curriculum) and the recipients of their work (that is, students). Consciousness does not always (indeed, maybe rarely) correspond to reality. But today, the stretch between historic prestige and guild professionalism on the one hand, and the current academic and workplace reality on the other, is so vast that the majority of active faculty are open to seeing themselves as primarily wage workers in need of a union.

CAN WE EVEN USE THE WORD "PROFESSIONALISM"?

If a guild mentality represents historic and, we would argue now, backward-looking aspects of professionalism, the question becomes: can a definition of professionalism be constructed that both reflects the current economic and political reality of most college and university faculty and strengthens their consciousness of their actual position?

The answer that we propose is that teaching is a craft, one that can also be called a professional skill, and essentially should carry with it the idea not of class superiority but of a set of skills, practices and ways of looking at oneself in the world that can be learned and taught. At the same time, teaching is a fundamentally political act that can be used for good or evil. This answer has been generally embraced within the faculty union movement since the formation of the AFT.

This is easier to "sell" for pre-K–12 teachers. It was carriers of this working-class craft ideology led by Margaret Healy in Chicago who formed a teachers' caucus within the NEA in the early 1900s.[4] These teachers went on to give birth to the "federation" movement that led to both the formation of local teacher unions in Chicago and elsewhere, and the national American Federation of Teachers in 1916 in explicit alliance with the American Federation of Labor, the predominant voice of the working class at that time.

More recently, Ben Rust, president of the California Federation of Teachers in the 1950s, is credited with the idea that teaching is neither a calling, like the ministry, nor a bundle of expert "guild" mysteries, but rather a craft that could be learned by any motivated, generally educated person within an open community of practice.[5]

It is therefore probably not an accident that the current wave of teacher militancy was first sparked by the Chicago Teachers Union (CTU/AFT Local 1) and its caucus, CORE (Caucus of Rank-and-File Educators), in 2012, in the very same city that had given birth to the idea of a working-class teacher movement a hundred years before. This strike embodied a perception of teachers as community professionals, whose work was paid for by public taxes and whose practitioners were fighting, not just for living wages, but for the conditions possible to deliver professionally adequate education to a student body that is predominantly working-class, and predominantly people of color. The union research report that laid out the strategy both for the union members and the broader public was titled *The Schools Chicago's Students Deserve.*[6]

Under the pressure of the rising teacher and higher-ed union movement, represented by the AFT, both the NEA and the AAUP have been forced to jettison their most egregious aspects of ideological elitism and strict guild consciousness. Over the decades, they have become, first in practice and then rhetorically, unions themselves and built links to the broader labor movement. The recent strikes in Los Angeles, Oakland, Denver, and at Wright State University in Ohio were all by unions that were affiliated with the NEA or the AAUP. Teachers, like nurses, social workers, and other female-dominated professions, have had to counter their image as self-sacrificing and show that collective action, including strikes that inconvenience the public, can be the necessary and effective way to get action on matters critical to their ability to do their jobs and therefore the public welfare.

DOES THAT GET US TO A USEFUL DEFINITION OF PROFESSIONALISM?

We would argue that this definition of professionalism would apply to academics in higher education as well. It can serve us as contingent faculty, our students and their communities, and the whole movement. It can help us avoid the consequences of what is sometimes called "the tenured gaze"—the view from higher up the elite mountain occupied by tenure-line minority. The tenured gaze wants to project itself as normative, and therefore cannot project to the public the fact and the consequences of having a contingent majority.

As guild professionalism divides us from the working-class majority, our definition of professionalism must identify ourselves as part of the working class. We have a craft pride that motivates us to fight for our students' learning and academic freedom. Whether it's called our professional knowledge or journeyman-level skills, it means knowing what is needed to do the job right at an adequate level that can be reproduced and sustained. Our professional concerns are about more than our individual job and career. This is not contradictory to recognizing our status as wage workers and part of the working class. We see professionalism as a collective endeavor, and we engage in collective action to further the goals of our profession in society. We learn the history of our profession and apply what we learn critically.

We struggle against our bosses to carve out space and time to cooperate with our colleagues, not compete with them. We value useful craft knowledge, even if there is no monetary reward for it at a particular moment. We try not to confuse the form, packaging and passport to wealth that some ideas possess with the real content of knowledge, skill and moral value of ideas. We have a moral commitment to our profession, and we can defend it logically outside the profession and enforce respect for it on the basis of social usefulness, not just power, wealth, or tradition.

Therefore, real professionalism can be a help to the best kind of unions and professional organizations because people are motivated not just by immediate material gain and security for themselves, but also by a legitimate, deeply felt concern for the social usefulness of our professional function, namely critical education of the next generation. We are not individual guild elitists drawing lines of exclusion, but are confident to unite collectively with everyone who can contribute to the cause of progressive critical education, regardless of their status, class origins, race, gender, or occupation.

APPLYING THIS CONCEPT OF PROFESSIONALISM IN THE CSUS

In practice, how have contingent faculty applied these ideas usefully? Again, we will use an example from the CFA and the Lecturers' struggle.

One question the Lecturers faced very early, even before the Higher Education Employer-Employee Relations Act, 1979 (HEERA) was passed, was whether or not to use energy to fight for access to the various faculty senates, the professional shared governance body in the CSU and many other systems. Likewise, should they fight for access to the other lower-level shared governance bodies such as department meetings? Lecturers were historically denied participation in these bodies. In a 1979 national AFT survey of local union presidents, only 30 percent responded that non-tenure-line faculty should have full academic rights.[7] Today, in the CSUs, that percentage varies by campus.

This question is full of practical contradictions. On the one hand, participation in both high- and low-level shared governance bodies means contributing one's professional knowledge to a critical discussion. It can be experienced as an honor. It can appeal to the pleasure of working side-by-side with members of the tenure-line elite and management. On the other hand, participation carries risks: If a contingent participates and speaks up from the perspective of the Lecturer majority, that person is likely to be marked as a troublemaker for stirring up concerns that were not already on the agenda. Many contingents who are allowed to participate in these bodies feel intimidated when they arrive and therefore do not dare speak from their own professional knowledge.

The converse of this is often raised by tenure-line faculty, who object to allowing contingents in shared governance bodies. They say that given our lack of job security and dependence on the whims of bosses and supervisors, allowing us into these bodies risks having them function as a catspaw for management, against faculty interests, because contingents will be afraid to vote against their immediate supervisors. A similar argument was made against votes for women, which similarly underestimated the capacity for independent agency on the part of the disempowered group. This issue is dealt with in detail in the *AAUP Statement on Contingent Faculty in Shared Governance Bodies*, which ends up recommending full participation except on tenure and promotion decisions.[8]

Given that these senate-type bodies, in this day and age, are usually only advisory to management and boards of trustees, should we use our energies as contingents to fight our way to gaining a voice and vote in these bodies? Can we see in them not just the vestiges of historic feudal perquisites of professors, but rather the seed of a system of workers' control of a workplace?

The alternative that is often posed has been that only unionization followed by binding collective contracts can make these venues safe for honest participation.[9] Then a contingent with a union who braves the tenured gaze and speaks out at a senate meeting might have protection through the contract, in case being identified as a troublemaker leads to retaliation.

John Hess said that he changed his mind on this question in the course of the struggle. Originally, when Lecturers suggested spending time trying to get access to the Senate, he argued against it, saying that the Senate was just sandbox politics: "What you can win in the Senate is dog shit—they have so little power. It's basically all advisory." Later, he decided not to discourage people from working in the Senate because it mobilized them:

> There's the issue, and then there's the capacity of the issue to mobilize and organize around. When they learned that the Senate is a dead end, they came and worked in the union where you could bargain things you could enforce. The movement is everything. You can be right and still be wrong. If people want to fight on something, you don't say no, you join them.

The impulse to fight for a place in the professional shared governance structure comes partly from a sense of equality that should be encouraged. John said that was at least part of the motivation of people who wanted to fight in that arena, even though that arena was in fact sandbox politics. A substantial portion of the contingent faculty and some of the most serious and conscientious of them were largely motivated by these sorts of professional concerns. Given this, and given that numbers of contingents were willing to fight on this terrain, any organizer of contingent faculty needs to be in there with them.

14
How Does It Feel?

All social reproduction work, which is what teaching is, entails emotional labor. Many contingents stick with their work for years primarily because of the profoundly positive emotional aspects of it. However, these positive aspects are always undercut by the conditions of our jobs. Organizers need to recognize both sides: the deeply satisfying and the painfully disappointing. The work of an organizer is to acknowledge the truth of people's emotional life while shifting it away from individual complaining into collective action, to move another grain of sand on the scales from fear and fatalism to the side of courage and hope.[1]

When John Hess went to his first meeting of the Lecturers' Council, after his return to California, he reported that what he observed there was just endless complaining, "pissing and moaning." The reality is that when contingents are first organizing at an institution where there has not been effective organizing in the past, or in their experience, the first wave of activism involves what may seem like endless rehearsals of individual humiliations, disappointments and abuse. A circle of ten to twenty contingents can engage in hour after hour of venting, telling complicated stories of unfair treatment. The essence of these stories is that the pain is unresolved; it just goes on and on.

The abuse is real, and airing it collectively confirms this. Therefore an organizer faces a problem: at the same time that you are trying to get people to fight for themselves, you are asking them to look beyond their personal experience and see the collective struggle. For many contingents, this is hard. To begin with, establishing the legitimacy of fighting for ourselves is not easy. We may be ready to fight on behalf of other people such as striking workers in other countries, immigrants, or communities of color in the US, and even our students—but not for ourselves (even if we are ourselves immigrants or people of color). Many of us still see ourselves as members of a privileged elite, floating intellectuals temporarily and unjustly shunted into precarious low-wage employment.

Much of what people do and think politically is not primarily based on intellect or "facts" but instead on their prior experiences and how they *feel* about their situation, their relationships with other people in that situation, and the collective struggle. Therefore the organizer has to construct collective experiences—for example, social gatherings, poetry readings and birthday parties—to introduce people to each other on the basis of common interests outside union activities. These may not be directly campaign-related, like house calling or phone banking, but they can work to build the social relationships of the group. The organizer has to trust people to let these relationships develop in their own way and give them time. This is real social capital that is collectively created and can be drawn upon in the future. It can work as an antidote to fatalism, fear, depression and anxiety: better to be angry than depressed, better to take action than be passive and despairing. Passivity and despair are emotions that come from being isolated and hopeless.

We should not forget to mention the powerful feeling of joy at finding "a home" in a union where one can be oneself and be honestly respected for one's work.

Here are some questions that contingents ask that sketch out the emotional terrain that an organizer has to work with.

The problem of catastrophic self-doubt

The stark contrast between the reality and the ideals of academic life can create a state of cognitive dissonance that requires significant emotional effort to sustain. Many of us, in order to avoid the pain of that dissonance, maintain a kind of double consciousness. On the one hand, we know that our working conditions are untenable. On the other, we know the significance of our work and what is needed to do it right. In the space between what is real and what is necessary, distorted beliefs can develop. For example, some of us who work as contingents convince ourselves that when the next job opening is announced, or the next tenured professor in our department dies, or when new funding appears to be flowing, we will be the one hired—"They have to, I deserve it, I'm the most qualified, I've been here the longest, I'm next in line, there's nobody else, etc." In fact, they don't have to. Delusional reasoning leads to behaviors that are counter-productive and sometimes disastrous. The structure of fantasy can collapse at the worst possible moment, at a faculty meeting or in a job interview. The dam breaks and unless checked by a terrible effort, unspeakable thoughts flow out, often with anger.

The psychological resources that sustain this double consciousness, and the awareness of the delicacy of the structure of fantasy, creates an abyss of self-doubt and self-questioning. The positive feedback loop that normally goes with doing a job year after year, even if it's only done moderately well, is absent. This is more serious than a mere lack of self-confidence about whether or not one is a good teacher, researcher or colleague. One can know intellectually that one is good at one's job and at the same time lack the confidence that should come with that.

If the individual tries to deal with this state of mind alone, things do not get better with time. However, giving up the belief that one is guaranteed the next tenure-line position and accepting the reality that this is a collective, not an individual problem, can be difficult. An organizer must respond to this resistance with patience.

Am I an impostor?

The "impostor syndrome" is usually something that is associated with starting a new job. A new hire in a better job with more responsibilities may feel that they have been hired by mistake and will soon be exposed. In the case of contingent faculty, the "impostor" syndrome is part of the cognitive dissonance between the high status of the job that we are supposed to do—and that in fact we are doing—and the low status that we occupy as contingents. Students and other campus workers, from custodians to librarians, call us "professor" and treat us with respect, while our department chairs and tenure-line colleagues treat us as here-today, gone-tomorrow disposable "units of flexibility," as one manager cheerfully explained. When this cognitive dissonance grinds along in someone's unconscious, it can surface abruptly and inappropriately as anger or embarrassment.

John Hess captured this when he said that none of us—no contingent—is "more than three seconds away from total humiliation." Walking down a hallway, having just taught an excellent class or graded a set of thoughtful, exciting exams, one can cross paths with a manager, who says, "Oh, by the way, there's no place for you on the schedule for next semester. Sorry!"

Made conscious, anger can be a source of energy. It is also possible to sharpen the dissonance into a joke. But confronting the reality is not easy. A contingent may stand in front of the mirror every morning and ask, "Am I really a college professor? Or am I just doing this until something better comes along?"

What can I do about stress?

Stress is a health hazard that is recognized by the Occupational Safety and Health Administration (OSHA). For contingents, the impact of the physical and psychological stress of our working conditions exacts a toll over the years that shows up in illnesses ranging from musculoskeletal disorders to allergies, auto-immune diseases, chronic pain, and depression and persistent anger. A faculty member with a desk and an office can get their university to provide an ergonomically correct desk and chair. The contingent faculty member cannot, and is likely to be driving many miles each day and grading papers in the front seat of their car in a parking lot, or at a Starbucks, and sooner or later will bear the marks of living in a constant emergency. Stress is not just about one's job: the degraded conditions of the job drag down the conditions of the rest of one's life, as the uncertain paycheck drains the bank account, postpones car repairs, and racks up penalties or interest payments. Symptoms for which someone who can take sick leave would visit a doctor, are postponed until vacation, if ever. All of us know someone for whom this had fatal consequences. A close friend of the authors in Chicago, the feminist and IWW leader Penny Pixler, worked as a substitute teacher in the Chicago schools for decades, lacking health insurance until the day she aged into Medicare at 65. Upon taking advantage of her insurance, she found that she had advanced breast cancer, which killed her only a few years later.

The uncertain class timetable also makes it difficult to schedule child-care, conflicts with other job possibilities, and often completely eliminates exercise and recreation. Cumulatively, the stress of working as a contingent can probably be best understood as causing PTSD.

MANAGING THIS TERRAIN TO ORGANIZE
FOR COLLECTIVE ACTION

Listening

John Hess got good at a technique that is harder than it sounds: listening. Not just to a brief statement but for an extended period of time: twenty minutes or half an hour. He used this in his organizing in the CSUs and also brought it to a COCAL Conference in 2006 and asked participants to practice listening. Many people have never sat and listened to someone else talk for that amount of time, uninterrupted. Many people have never received that kind of attention before. From the organizer's point of view, this exercise is more than just a gift given to someone who needs it; it is efficient, in terms of the amount of information the organizer gathers.

When he used listening as an exercise, John would use an open-ended prompt like "How did you get here?" as a way of suggesting that this was a chance for speaker to tell their story.

Turning pain into laughter

At the Lecturers' Council meetings, John Hess and Elizabeth Hoffman, who was the Lecturer vice president of the CFA overall, taught a game in which delegates would create a list of insults. The list would include the actual words or gestures that were insulting, the name of the person who did it, the situation, and the meaning of the insult as experienced by its target. The emotional reactions would start as private shame; the laughter of the group made them public. Each insult and its supporting material would then be analyzed in a collective discussion. What was the actual content? What was really being said? Who actually uttered it? What did the whole list of insults reveal? This process is somewhat similar to the "inoculation" process learned in organizing trainings.[2] Each delegate would then go back to their campus and teach the game to faculty there.

Another game used by other groups of contingents was to see who had the highest number of parking permits from different campuses or miles driven on one day? Who had the highest number of different textbooks for the same class, different physical locations or institutions, different subjects, total classes? What was the most clueless thing a department chair had ever said to them, or the strangest place they had ever met with a student? This competition could also let people confess mistakes: how many had driven to the wrong town to teach a class? How many had carried the wrong class bag to a class?

Lessons from the Ruckus Society

The story of how John Hess invited the Ruckus Society to come to a Lecturers' Council meeting is told in Chapter 3.[3] One lesson from that meeting was that people learned how to be systematic and disciplined in preparing to take their anger into the public arena—how to heat up this emotion, dramatize it, and take advantage of both legal and physical space (laws, rights and the geography of streets and buildings) to express it politically. This activity valued and validated their anger. It helped people shake off the fears and hopelessness that so many were experiencing. It also gave them training in very physical collective direct action (such as locking arms) that formed a basis of experience that they could use later when solidarity was threatened.

USEFUL CONCEPTS FROM OTHER MOVEMENTS

The social justice movements of the last seventy years, all of which push towards equality, have generated concepts that are useful both in appreciating the emotional terrain of the contingent faculty struggle and in figuring out how to work constructively with it. These are concepts from the civil rights, gay rights and women's movements (and even the labor movement), among others. All of these concepts have been attacked, incidentally, by the political right wing. Our adoption of these concepts into an occupation- and class-based condition, namely second-tier faculty, is a useful extension of them, not, we will argue, a wrongful misappropriation of them.

The concept of "microagressions" has come out of the anti-racist struggle and speaks to a type of insult that by itself might seem trivial but is woven into a whole discourse or set of behaviors that signal, sometimes subtly, the superiority of one person or group, and disrespect and contempt for the other. Microagressions form a tapestry of signs that the recipient recognizes but cannot easily respond to.

A contingent historian, for example, was jokingly told upon being hired for his third consecutive year that he had better watch out, "because after three years as an adjunct you are used goods." At a meeting with other contingents, the historian laughed along with the joke. He also described a sinking feeling of shame that he did not understand until later, when he realized that "used goods" means a woman who has lost her virginity and therefore will have a hard time finding someone to marry her. The insult had a double impact: his boss is calling him a woman, and a damaged one at that.[4] In another case, a tenure-line faculty person kindly advised a contingent that "one really has to be of independent means in order to pursue the social sciences or humanities."

A familiar microaggression is the use of the term "faculty" to mean only tenure-line, not contingent, faculty. Elizabeth Hoffman tells the story of noting in the contract that "faculty" voted in the choice of whom to recommend to the dean as department chair. She pointed this out and tried to vote. This was very divisive:

> Someone said to me, "Why not let students vote?" One guy said, "We should let my dog vote." Another one actually mentioned a three-fifths vote. They were going to have different colored ballots. Another guy said, "You don't understand—if we give the Lecturers a vote they'll vote their self-interest!"

But since Lecturers were legally faculty, the union grieved to allow Lecturers to vote for department chair recommendations, and won.

Coming "out of the closet" is a concept that started being widely used in the gay rights movement back in the 1960s. Contingents share with LGBTQ people the ability to hide our difference from what most others expect us to be. Just as LGBTQ people can hide, if they want, behind the assumption of heterosexuality, we can hide behind the popular assumption of "professordom." In both cases, the price of "staying in the closet" and hiding the reality of our lives from those we interact with professionally, casually, even in our families, is very high. The liberatory step of coming out of the closet is a very useful concept for us, just as it has been for the gay rights movement and also for those on the margins of racial culture in the US—that is, those who could pass for white instead of Black, Asian, Latinx, or Native American. Students are almost always surprised to find out that a professor is a contingent, and sympathetic when they find out.

Another concept is "the personal is political," which comes from the second-wave women's movement in the 1960s and 1970s. This concept demands that we try to connect our personal stories to the big picture of the political economy of higher education, and asserts that these stories belong in this picture. This entails coming out of the closet, of course, and making sense of the contradictions that cause the cognitive dissonance that over time wearies and damages us.

The environmental movement has given us the concept of sustainability, meaning a state in which living things are in equilibrium with their world, whether they are plants, animals, or the creation of human beings. The usefulness of that concept to us is that having a casualized faculty with all the stresses that go with it is not ultimately sustainable, for the faculty, for the education system, or for the students. We also talk about what a sustainable life as a faculty would look like in our "Blue Sky" agenda.

The civil rights movement has given us so many ways to think about fighting for equality from a subaltern position that one has to pick and choose. One phrase in common usage is "the back of the bus." For contingents this phrase, used in a bitter tone, can refer to the gang offices, locked supply or photocopy rooms, lack of access to departmental decision making, and other realities that are felt as wounds.

This is not to equate the US history of racial oppression and violence with the experience of contingent faculty. We are talking here about how to understand the emotional terrain. Frederick Douglass, in his 1852 speech, *The Meaning of the Fourth of July for the Negro*, speaks eloquently to the emotional experience of being a second-class citizen. He begins "Oppression makes a wise man mad …" and, later in the speech, continues: "The blessings in which you, this day, rejoice, are not enjoyed in common. The rich inheritance of justice, liberty, prosperity and independence, bequeathed by your fathers is shared by you, not by me."[5]

The emotional damage done by second-class status is severe enough that someone holding second-class status, whether because of race, gender, nationality, age—any of the named categories of discrimination or any others, for that matter—can recognize Douglass's furious logic and feel moved.

We should never underestimate the power of inspiration that can come from other related movements for liberation. Just as the anti-slavery movement and the Haitian Revolution were inspired by the US and French

Revolutions, all the movements of the 1960s and 1970s were inspired by the civil rights movement. The emotional aspects of these can flow over into later movements in unexpected and powerful ways.

PTSD and moral injury

Returning veterans, first from Vietnam and then from the successive US wars, have made the term "PTSD" part of common discourse. Post-traumatic stress disorder is both psychological and physical and can last many years. Contingents recognize descriptions of PTSD in their own patterns of obsession and anxiety, and use the concept to give a name to a condition that comes from suffering or witnessing a terrible experience. "Moral injury" is a related concept, with the difference that the injury at the heart of *moral* injury is witnessing something that is grossly unfair happen to someone else without being able to prevent it.[6]

As a result of research into the experiences of PTSD-afflicted veterans, treatments have been developed, but they tend to be individual-based (except for group therapy), like de-sensitization or medication, and are not likely to include an analysis of the political economy of war. However, other research has found that joining collective action to address the social causes of a problem (for example, organizing against the stigma of mental illness) is in itself therapeutic. In fact, overall, the opportunity to participate in collective action has a positive effect on people with anxiety or depression. In a word, the way to feel better is to join the movement. The experiences of those in groups such as Vietnam Veterans Against the War and Iraq Veterans Against the War (now renamed About Face) confirm this.

A safe space

There was a moment at a COCAL conference organizing workshop in 2006, when John Hess was conducting a workshop and the room was full of people including staffers and union leaders, some of whom were of course tenured or tenure-line faculty. Unexpectedly, John Hess asked everyone who was not actually a contingent to leave the meeting room. His request, made politely but firmly, was a shock both to the tenure-line participants, who were accustomed to having full access to conference proceedings, to union staff who considered themselves allies, and to the contingent participants who were accustomed to being surveiled, if not directly observed at all times. But that moment, during which the tenure-line faculty and union staff got up, gathered their things, and filed out, was a powerful transitional moment.

That moment said many things. One was about the right of contingents to take over a space for themselves. Another was that this would be a safe space: a space shared by people who had a common concern and a commitment to do something about it. A third was that along with this space itself came some power which could be claimed by everyone who was there. That moment declared: there are interests and concerns that are particular to us and we have to organize ourselves, separately, in order to win them. No one can do it for us. It was an example of modeling an element of the Inside/Outside strategy. It was also a reversal of something that had been taken to be irreversible. It had something in common with the moment when John Hess realized, thinking of tenure-line faculty, "They have nothing to teach us; we can teach them."

A WARNING

As organizers, we sometimes have to confront the fact that one of our people is so damaged by their life as a contingent that they can't be worked with in the collective struggle. They might recover with time, but for now, they cannot help. Hence, these people need to be counseled out of the struggle or organization because they can in fact do real damage. Sometimes they may need to be counseled out of the profession itself, because they are so damaged that their usefulness as a teacher may be outweighed by the damage they have suffered psychologically. These are hard decisions to make, but they are necessary for the organizer to confront to avoid the damage some of these colleagues can do to the collective struggle and the group spirit. Nearly every one of us has snapped under the ongoing stress of our working conditions and gone over-the-top for a spell, but some of us spend more time there than others.

15
"Is this legal?"

This chapter is less self-reflective than the others, and offers more explanation and information. We need to respond to the situation of people who are thinking about organizing or who are planning actions in support of contingent faculty. Planning of this sort can come to a dead stop when they ask or are asked, "Is that legal?" Then their minds turn to the question of the law: what does it say? What does it allow them to do? Without information, often, that's the end of the story.

The first thing to know about the law is that it is written, unwritten, and rewritten by human beings. The Second Law of Thermodynamics may be a fact of nature, but the laws we are talking about are not. Like collective bargaining contracts, human-made laws reflect the power relationships of the moment. Before 1900, there was no positive provision for collective bargaining in the US. That situation altered not primarily because of a change of heart among elected leaders, government agencies, or administrators, but because of organizing by workers on the ground. Workers formed unions, but without legal authorization. They then demanded the legal right to form unions that would be recognized and with which employers would have to bargain. This demand had to be backed up with direct action, up to and including strikes. Over a period of decades, this right had to be fought for all over again. This means that unions in education and in every other sector pre-date the law and have brought the law into being by their actions, not the other way around. The struggle makes the law, the law does not make the struggle.

So unions came first, then the laws that regularized them. They are, however, among the least well enforced of all laws, all over the world. Their enforcement in practice comes from the same source that the laws came from in the first place: namely, workers who watchdog and report violations, and direct action by workers that gives meaning to them. Perhaps the only other laws as poorly enforced are those against domestic violence, and for the same reasons.

Of course, another time when people ask, "Is that legal?" they are really asking, can they do that to me? Can my employer really fire me, discipline me, promise me something and then not do it, etc., etc.? The answer in this case is almost always "Yes—unless you have a union." In the US, we are all employees-at-will, meaning the employer's will, and with very few exceptions the employer can do anything they want.[1] The exceptions are where the employer might violate some federal or state law, in which case the answer is, unfortunately, likely to be "What the employer has done is actually illegal and prohibited, but very hard to fight unless you have a union." But what we are concerned with here is whether what we are planning to do is legal or not. In either case, what really matters is how much power you have.

PUBLIC-SECTOR VERSUS PRIVATE-SECTOR LABOR LAW IN HIGHER EDUCATION: WHAT LAW GOVERNS US?

The next thing to know is the difference between the public and private sectors. In the US, a line is drawn between directly tax-supported and publicly governed "public" institutions, and privately owned and controlled institutions, either for-profit or nonprofit. It is important to know which kind of institution you are in because the laws are different. This may not correspond to what makes sense, given that all of higher education depends on public money, and usually heavily on public money. The only exceptions are a few elite universities that have very large endowments and rich alumni, but even these get some public funding—for research, capital projects, student aid and government grants. For state-controlled higher education, money comes as direct appropriations from legislatures and, in the case of community colleges, from property taxes. So in a very real sense, all of higher education should be viewed as "a public good," but only some of it is governed by public agencies that are in any way answerable to the public through democratic electoral processes. That's the difference between "public" and "private."

This distinction and this terminology is unique to the United States, which is the only country that has had a very large, private, nonprofit higher-education sector. This goes back to the 1890s, the "robber baron" period, when the Carnegie Foundation and the other corporate-created foundations were set up. Before that, the only "nonprofit" and tax-exempt organizations were churches, from which much of private higher education originated.

The for-profit institutions, places like the University of Phoenix, DeVry, Grand Canyon University and Kaplan, which are a recent phenomenon, are obviously part of the private sector. There are also a number of institutions whose private/public status is unclear. There has been litigation about whether they are covered by state collective bargaining laws. Perhaps the largest examples of these are the "state-related" institutions in Pennsylvania: Temple, Penn State, Pitt and Lincoln.

PRIVATE-SECTOR LABOR LAW IN HIGHER EDUCATION

Laws enabling collective bargaining are one of the key issues that we have fought for. In addition, we have fought for the protections of the other employment laws that exist at the state and federal level. In the private sector, contingent faculty are almost universally acknowledged to have bargaining rights and mandatory recognition of a demonstrated majority in a proposed bargaining unit under the National Labor Relations Act. The NLRA established the National Labor Relations Board to administer and adjudicate under the law. It also listed violations of the law, known as unfair labor practices (ULPs), specifying that these might be committed by either side. These charges are adjudicated by the NLRB.

This has been settled law since the 1970s, when college and university employees, and employees of other nonprofits, were brought under the NLRA. This status has been reaffirmed repeatedly and especially in the last seven years since the Service Employees (SEIU) has been actively organizing contingent faculty in the private sector under the campaigns first called "Adjunct Action" and then "Faculty Forward."

The "management" challenge

There have been two major legal challenges to our right to collective bargaining in the private sector. At the time of writing, these have been settled in favor of contingent faculty's right to organize. However, given the composition of the NLRB and the Supreme Court, these decisions could always be reversed. In fact, decisions on graduate employee bargaining rights have already been reversed repeatedly.

The two challenges have revolved around first (and least important for contingent faculty), the 1980 *Yeshiva* decision, where the Supreme Court overruled the NLRB and judged that full-time regular tenure-track faculty in private-sector institutions (with the example being Yeshiva University in New York) are in fact "managers" because they exercise decision-making power within the institution. As managers, they do not fall under the rubric of the NLRB, and therefore the employer is not required to bargain with a demonstrated majority-representing organization. However, the *Yeshiva* decision did not forbid collective bargaining, so a few full-time faculty unions in the private sector remain organized and recognized by the employer.

Recently, in the case of the attempt of the SEIU to organize full-time and part-time contingent faculty at Pacific Lutheran University in Seattle, the NLRB decided that full-time non-tenure-line faculty were not covered by *Yeshiva*. They denied the university's objection to recognize the bargaining unit on that basis. In its Pacific Lutheran decision, the NLRB laid out some new and more restrictive criteria to be used in applying the *Yeshiva* decision. In general, that means that probably, with these new rules and criteria, combined with the fact that full-time tenure-line faculty have been losing shared governance and managerial prerogatives in the years since 1980, a large percentage if not the majority of tenure-line faculty are now conceivably unionizable under the NRLA. Certainly, for all non-tenure-line faculty—who do not have permanence of employment, to say nothing of adjuncts—*Yeshiva* should not stand in the way of their being organized under the NLRA.[2]

The "religious" challenge

A second aspect of the Pacific Lutheran NLRB decision concerned the objection by the university management that it should not be covered by the NLRA on the basis that it was a religious institution. A great many colleges and universities (though not as many as a hundred years ago) are affiliated with religions and religious organizations. In recent years, many of them have challenged the right of their faculties to unionize under the NLRA on the basis that the Act provides for the voluntary exclusion of religious organizations from coverage. The question is whether these colleges and universities are "religious organizations" in the meaning of the Act and whether the work of their workers is religious work. The NLRB decided this case and subsequently a number of others, including Manhattanville College in New York, St. Xavier in Chicago, Carroll College in Wisconsin, Seattle University and Pacific Lutheran in Seattle, and Duquesne in Pittsburgh.

The NLRB ruled that to qualify for the religious exemption, the college had to demonstrate that the work the faculty was doing was the propagation of religion, and that they were explicitly hired to do that as part of their job. Since none of these colleges could demonstrate that—except for the religion and theology teachers, who were excluded deliberately—this decision opened the door for organizing in a number of religious institutions. It should be noted that some religious institutions have broken ranks with this position and have accepted unionization: Georgetown, for example, and in a previous generation, the University of San Francisco (both Jesuit schools). Debate is continuing to rage within the Catholic community, especially over the argued violation of Catholic social teachings that resistance to unionization represents. An entire organization, the Catholic Labor Network, has been formed to combat opposition to unionization.[3] They are also dealing with Catholic hospitals and social service agencies that have taken this position.

In the current political situation, the Pacific Lutheran decision stands a strong possibility of being reversed either at the NLRB or at the Supreme Court levels, as does the current graduate employee decision. But short of reversal, private-sector contingent faculty in both the for-profit and non-profit sectors have a clear legal route to unionization. This does not mean that employers will not raise *Yeshiva*, or the religious objection, or other spurious claims, in an attempt to discourage and confuse organizing, as they have in the past.

One final wrinkle is that current NLRB decisions seem to allow units of one department or one school within a university to organize. That has been applied in some graduate student employee units, so-called "micro units." With the exception of "carve-outs" of some professional schools like medicine or law within a university, this has not commonly been the case within units of non-tenure-line faculty, but we may see attempts along this line in the future if the legal possibility of micro units continues to be in effect.

PUBLIC-SECTOR LABOR LAW IN HIGHER EDUCATION

For those who are teaching in a public-sector institution, a different set of laws applies. In the United States, federal labor law, as compared to state-level law, applies only to the private sector, for-profit or nonprofit, and not the public sector. Starting in the late 1950s, states enacted laws regulating public-sector employment in their jurisdictions. Usually, this was under the pressure of strikes or threatened disruptions by unions seeking to have the same sort of mandatory recognition and collective bargaining rights that exist in the private sector under the NLRA. At this writing, approximately half the states have some degree of public-sector collective bargaining allowed by statute. Most of these laws are based to some degree on the NLRA. A few are better for workers, in that they have been interpreted more broadly and allow broader coverage of workers. Most, however, are more narrow both in scope and in the regulation of what workers can do to promote their interests. Many, for example, include prohibitions against striking.

State-level public-sector labor laws all mirror the National Labor Relations Act in that they include a list of acts that are specific violations of labor law. As under the NLRA, these are called unfair labor practices. Even if workers have an existing contract with a no-strike clause, they often have more rights to respond collectively to an unfair labor practice committed by an employer than they have to respond in an economic dispute. Union actions, like picketing, taking a strike vote, or building a credible strike threat, and even including striking, are not prohibited if they are done in response to an employer's unfair labor practice (ULP).

Unlike federal labor law, state laws have a myriad of forms, coverages, jurisdictions and titles that are then interpreted by the various state courts. Only very rarely, if a federal constitutional question is raised, do these get to the federal courts. This means that the generalizations we can make about public-sector higher-education workers' rights are limited. It is also worth adding that, with few exceptions, the present trends in public-sector labor law are almost all negative for workers.

Examples include the restriction of the scope of bargaining to only wages, or even more limiting, wages within the inflation rate (as in Wisconsin or Iowa); a requirement that unions recertify by election annually (also Wisconsin and Iowa); severe restrictions on union political activity, or, in the recent *Janus vs AFSCME* case,[4] restrictions on the unions' ability to bargain the form of union security referred to as "agency shop" or "fair share." Fair share had been legally bargainable in California, and in 1999 the unions who supported Governor Gray Davis pressured him to make it mandatory for all state and local public employees. Since the *Janus* Supreme Court decision of 2018, fair share in the public sector is totally gone; it cannot even be bargained. This prohibition against fair share is what is called by conservatives, employers and enemies of unions, "right to work," meaning in fact the right to benefit from union-bargained protections without having paid union dues.

That being said, since education is still mostly a public-sector function in the US, most of our higher-ed bargaining agents and unions are in the public sector. Historically, because of the *Yeshiva* decision in the private sector, this trend has been even more pronounced: the majority of unionized workers in higher education are in the public sector. This is also one of the few sectors of the overall labor force where the percentage of unionized workers has been growing. This is due primarily to organizing among non-tenure-line faculty by the SEIU, UAW, the Steelworkers and CWA, and by the traditional education unions—the AFT, NEA and AAUP.

ENABLING LAWS ARE THE PRODUCT OF DIRECT WORKER ACTION

What can we say in general about the legal terrain for public-sector higher-education faculty? Rather than attempt to talk about particular cases, some points of principle might be most useful.[5]

First, just as is the case with the federal labor law, the various state labor laws are almost all the product of direct worker action and subsequent lobbying. But they have never been the result of lobbying by itself.[6] The people who carried out these direct actions undoubtedly asked the question, "Is this legal?" In the cases where, technically, it was not, people had to act despite this, in the hope that the law might be changed through their action. The credible threat of disruption has always been a factor. In dealing with state labor laws on this matter, this is a very important lesson because state labor laws are often easier to change through direct action, followed by legislation, court decisions, or administrative actions, than laws at the federal level.

Second, contingent faculty are often poorly protected or even overlooked by state labor laws. The best laws make no distinction between part-time, temporary, and full-time or permanent workers in higher education or elsewhere. This is the case in California. But in many cases, many or even most part-time temporary faculty are excluded from labor law protection, either explicitly in the statutory language or through past labor board decisions. In many jurisdictions, legal language makes distinctions that do not correspond to the real-life experiences of workers. For instance, some non-tenure-line faculty might be defined as "regular" and others as "casual," a distinction that is in most cases deceptive and not useful. In the case of Illinois, contingent faculty who worked less than a certain number of units were at one point excluded from representation (this exclusion was changed in 2000 through lobbying and organizing in the sector).

Third, many of these state labor laws exhibit some of these same problems with regard to graduate employees, who are of course a group that heavily overlaps with other contingent faculty—sometimes even being the same person at the same time. These have also given rise to direct action pressure on state governments, even strikes, lawsuits, lobbying and legislation, in which the general trend has been victory at long last. "Long" is, however, measured by years or even decades.

Fourth, most states that have collective bargaining laws have established some sort of labor board to enforce the laws and adjudicate disputes under them, either directly or indirectly through supervision. Most have some structure to set or confirm bargaining units and judge unfair labor practices as their laws define them. In most cases, state labor boards tend to look for advisory purposes at the National Labor Relations Board, though they are not bound by what the NLRB has decided. With regard to contingent faculty, this means that that pattern of unit determination has varied from state to state, from subsector to subsector (community colleges, universities), and over time, within a state and even within the same labor board. Most states, though, have generally adhered to the NLRB's long-standing precedent that if the two sides can agree on the bargaining unit, it does not have to be the "best" or "most rational" bargaining unit. It only has to be an *adequate* bargaining unit that does not include within it people with inherent conflicts of interest. Put another way, it should reflect an effective "community of interest." In cases where the parties cannot agree, the labor board must take decisive action and define the unit. Since most labor boards and administrative law judges are neither familiar with higher ed nor with the changes in academic staffing in the recent past, these decisions are often slow, confusing and ill-informed, even with the informative briefs filed by the parties.

It is really impossible to make a generalization about how legal bargaining units have shaken down at the state level, with one exception: higher-education managers have come to a general position opposing units that are combined tenure-line and contingent faculty. This was not always the case. However, it does seem to be their dominant position in legal filings in recent years.

The best bargaining unit for contingent faculty is a matter still of great debate within the contingent faculty and faculty union movement generally. Should a bargaining unit include just contingent, or contingent and tenure-line faculty? Or, slicing it even finer, should it include just full-time contingent faculty (however that is defined)? However, it does appear that management would rather bargain with separate units than with a potentially more powerful, united faculty. Managers seem to have come to the conclusion that they are in a stronger position when dealing with separate units and faculty are in a stronger position dealing with combined units. Their judgment in this matter is worth our taking seriously.

OTHER ASPECTS OF STATE-LEVEL LAW

Besides laws covering collective bargaining, laws that are important to us are those governing access to public pension plans, unemployment insurance, sick leave, any form of state disability insurance, and Social Security, and coverage by state-level OSHA, since federal OSHA does not cover the public sector at all. Some states have their own OSHA plans for the public sector. Also, some states have passed statutes regulating such issues as the maximum hours (or credits) that a contingent faculty member can teach at a specific institutions (such as the California community college 67 percent law[7]). Other states have legislatively mandated pay for office hours, coverage for health insurance, and even some degree of job security. In many cases, these mandates have established that the matters should be bargained locally.

SO WHAT ABOUT THE REAL QUESTION: "IS THIS LEGAL?"

The purpose of this chapter is to prevent people who are considering organizing from stopping dead when they come to the question of whether what they want to do is legal or not. We should have made it clear by now that what is legal or illegal differs not just from place to place, from time to time, and from sector to sector, but also, and more important, what is legal is itself the product of organizing. What is not legal today can be made legal tomorrow if enough people organize to make it happen that way. The law generally changes after enough people break it. Therefore, there is no "illegal strike." There are only strikes that are not strong enough.[8]

The question, "Is it legal?" needs a response that acknowledges the emotional terrain. The person who is asking it may have been invited to a meeting off-campus. The meeting may feel secret, and in fact, *should* be kept secret from managers, if it is an early organizing committee meeting. The other people at the meeting may be people that the questioner doesn't know, or perhaps has seen speaking out in a way that marks them as willing to take risks. The whole feel of the meeting may suggest something that will get participants in trouble. Popular and media depictions of labor history tend to focus on strikes, violence and police beatings. So the person who is asking if what the group is up to is legal is also asking for reassurance that goes beyond just saying, "Yes, organizing a union is legal in the United States." Or not, as the case may be!

Obviously, learning about the legal terrain within which you are working is of great importance. This probably means getting good local legal advice. That's true especially if you are in the public sector. This is often easier said than done, since the labor side bar in the public sector is often quite small and limited in many states. The nature of this advice should be a description of the legal terrain within which you are hoping to have some effect.

Another way to look at the law

The question "Is it legal?" is often asked out of fear that some action is going to get an individual in over their heads in legal trouble. These questions come up a lot during organizing, but also when the union or a group within it plans to form a caucus, endorse a candidate, leaflet, petition, picket, sickout, do a grade strike, etc. Responding with "Don't worry about it" or "The union will take care of it" is condescending and false, and undermines the trust of the person asking the question. An equally unhelpful response may come from a manager or tenure-line union leader who says, "Our lawyers say that's not legal."

A different way to respond is to say, "Let's look at the law itself." In this case, the law is often "the Act," the actual National Labor Relations Act. Written over eighty years ago, the language of the NLRA expresses a vision of parties negotiating with each other as equals. Of course the NLRA has been amended and weakened, and of course the politics of appointing the NLRB have hollowed out hopes of enforcement; in addition, public-sector labor law is somewhat different though largely based on the NLRA, and there are many other ways in which it does not in reality protect workers the way we want. However, that vision of equal negotiators is still embedded in the Act. Making sure that people who are interested in (or worried about) the legality of collective actions—"concerted activity"—can see and share that vision is important.

Looking at the law itself means either going online to look at it (www.nlrb. gov/how-we-work/national-labor-relations-act) or getting a paper copy, such as the one in the collection by Robert Schwartz.[9] Having a copy of the law itself to study and discuss can be a bit like having a copy of the Bible in one's own language instead of Latin: you get to see what the law actually said. Despite the ragged history of the implementation of this law in practice, the vision it paints has the ability to inspire. For example, in Section 157, which is the old Section 7 (and which people today refer to as Section 7 rights):

Right of employees as to organization, collective bargaining, etc.

Employees shall have the right to self-organization, to form, join or assist labor organizations, to bargain collectively through representatives of their own choosing, and to engage in other concerted activities for the purpose of collective bargaining or other mutual aid or protection and shall also have the right to refrain from any or all such activities except to the extent that such right shall be affected by an agreement requiring membership in a labor organization as a condition of employment as authorized in Section 158 (a)(3) ...

There is probably no union or organizing committee anywhere that would not benefit from sitting together and going through that paragraph to unpack individual phrases. What does "the right to self-organization" imply? What is included in "form, join or assist" labor organizations? We may think "bargain collectively" is obvious, but why doesn't it just say "bargain"? How about "representatives of their own choosing"? And probably the most important of all these powerful phrases is "concerted activities," especially when it is connected to "or other mutual aid or protection." And what about the verb "shall"? The last part, beginning with "shall also have the right to refrain" will trigger a brief history lesson about Taft-Hartley.[10]

Another place to look is at the list of unfair labor practices, Section 158 (the old Section 8), and includes forbidding actions by the employer, sometimes summarized as "intimidate, dominate, retaliate, and discriminate." In many cases, the person who asks "Is this legal?" has been illegally intimidated.

This language was tooled as carefully as the Book of Genesis and is deserving of an equally thorough and careful explication and analysis. The fact that it has been watered-down, amended, and weakened through lax enforcement does not mean that the original vision was unclear. The additional fact that this is still, despite everything, our overarching labor law after nearly ninety years, given everything else that has happened, is inspiring in itself. Anyone who asks anxiously if something involving concerted activity for mutual aid or protection is legal needs to spend some time with the text of the Act, preferably in a discussion group with other people.

Incidentally, this is the only law in the US Code that grants groups the right to collective—that is, concerted, as in acting in concert with others—activity that individuals do not have alone. All the rest of the US Code regulates individual activity, with the exception of conspiracy laws, which ban certain kinds of collective activity. This makes the NLRA the most radical provision in all of US legislative law.

16
What About Leftists?

Do we really need the radicals, or are they just a pain? What is the proper, needed, historic role of political radicals (leftists, socialists, communists, anarchists, syndicalists—pick a label) in the struggle?

We write this as open radicals and socialists ourselves. One thing that we have discovered in our own careers and our research, as well as in our participation in the movement, is that almost always when a group begins to come together to try to change their reality, at least one member of the group, no matter how few people belong to it, is a radical of some sort. If not a law of nature, it seems to be a strong tendency.

Take for example the general strikes in 1934 (in San Francisco, Minneapolis, Toledo, and the whole-industry general strikes in California agriculture and the textile industry). These sparked the passage of the National Labor Relations Act in 1935. Then there is the great sit-down strike in Flint, Michigan, against General Motors in 1937. That strike opened the door to the successful unionization of the mass production industries under the Congress of Industrial Organizations (CIO). It is not an accident that these were all stimulated and led by organized political radicals, communists and socialists of one sort or another.

"Organized" here means people who have a relationship with an organized radical group, such as being a member or working with such a group, not just an individual with radical ideas. In the examples above, the group played a role, providing advice, strategy, support, a space for discussion and debate, mobilizing solidarity from other workers and communities, including food, money, publicity and legal assistance. The strikers' connection to an organization was key. Each one of those strikes had activists and leaders who were also leaders in their Left organizations.

The Chicago teachers' strikes, in 2012 and 2019, were led by the Caucus of Rank-and-File Educators (CORE), the insurgent caucus that included political radicals of various stripes who had been elected into key positions in the CTU in 2010, as told in *How to Jumpstart Your Union*, from Labor Notes.[1] The recent successful teachers' strikes of 2018–19 in West Virginia and Arizona were likewise inspired and often led by explicit socialists. Eric Blanc, in his book *Red State Revolt* about the teacher strikes in West Virginia, Arizona and Oklahoma ascribes the more limited success of the Oklahoma strike to the absence of radical leftists in the effort.

So the questions for this chapter are: if it's true that radicals are necessary for organizing, why is it true? And how might people strategize to take maximum advantage of this reality and minimize its downside?

THE ABILITY TO IMAGINE AN ALTERNATIVE

It is axiomatic among organizers and activists that in order for people to fight to change anything, they have to be able to imagine something different. People cannot have hope for something different unless they can imagine something different. Otherwise, they are imprisoned by fatalism, which, combined with fears of retaliation and ostracism, succeeds in keeping most oppressed and exploited people quiescent most of the time.

In an early interview, John Hess, who called himself a socialist, said, "We are the ones who can imagine the future—imagine the future in broader terms: why we fight." This broader imagination, stemming usually from some sort of broader social and historical critique and awareness of what is going on in other movements, gives radicals of all sorts (socialists, communists, anarchists, Wobblies, members of any of a number of smaller groups) two things that are essential in any struggle. First is the courage to step forward when few others will, and second is at least the rudimentary tools of organizing to move from individual complaint to collective action. Without somebody with these tools and this courage, no organizing attempt or social struggle can effectively get off the ground and reproduce itself among the people affected.

Susan Meisenhelder noted that the open meeting at which members of the CFA debated pro and con the rejection of the tentative agreement "brought out a lot of old leftists." She doesn't explain what difference that made; she seems to assume we know. Indeed, it is sometimes a source of amusement among contingent faculty activists to note how many among us have PhD's in philosophy, a field that requires a practitioner to handle abstract concepts and imagine alternative realities at the same time. Philosophers also usually know something about Karl Marx, who was awarded his degree in Philosophy in 1841.

One such "old leftist" was Mario Savio. In Elizabeth Hoffman and John Hess's chapter "Organizing for Equality within the Two-Tier System" in Keith Hoeller's book *Equality for Contingent Faculty*, they describe what they call "the Mario Savio effect." Savio was a hero of the 1964 Free Speech Movement (FSM) at Berkeley, and later a Lecturer at Sonoma State (one of the CSUs). He was a delegate to the Lecturers' Council in the 1990s during the depressing period described by Susan Meisenhelder, when very little had been achieved for contingents through bargaining. Hoffman and Hess say:

> "There are a lot of Lecturers in the CSUs," Savio commented. Then he asked, "How many Lecturers are usually at the bargaining table?"
>
> "One," someone answered.
>
> "There always has to be at least two of you if anything is going to change," Savio said. That was a revelation to most of us in that room, but by the end of that Lecturers' Council meeting, we understood that getting more representation for Lecturers was part of a broader organizing strategy and we did ask for—and get—another Lecturer on the bargaining team, the beginning of significantly improved representation in the years to come.

Savio was able to identify a roadblock, visualize a simple alternative, and propose a concrete step that could be taken that was immediately do-able and likely to be successful. His suggestion that any organizer find a partner to start with echoes Nina Fendel's recommendation that John go up to Chico and get together with Jane, who was to become a partner with him in the work with Lecturers. Savio's ability to pinpoint a next step probably came from a broad education (he was going to be a priest, studied philosophy, but eventually went back and got an MA in math and taught math) plus experience in the civil rights movement, the formative experience of his generation, where he worked with the Student Non-violent Coordinating Committee (SNCC). One of the sayings in the civil rights movement was to "never be the only"—in this case, the only Black person in some situation. One Lecturer at a bargaining table was an "only."

John said, "Like the lone woman in a room full of men, or a lone Black in room full of whites, just one Lecturer on the bargaining team will probably feel more alienated than if they weren't there at all. If we were there in any force, we would clearly want to change how it functions."

The question is often asked, "How bad do things have to get before someone does something about it?" This is not the right question. Things can get extremely bad, but without the tools that radicals bring, change does not happen. Change will remain stillborn even if a campaign is well-resourced, planned and objectively necessary. Big treasuries, paid organizers, research departments and strategic communications campaigns are not what is key. The largest example of this, albeit negative, in the history of the US labor movement was the campaign called "Operation Dixie," an attempt to organize the South. This was in 1946–47, right after World War II, a time when radicals were being purged as the US hardened into the Cold War. The alternative that the left-led unions imagined was a biracial explicitly anti-racist labor movement in the South. Such an effort was led largely by Communists throughout the South in the 1930s and '40s.[2] But the white leadership of the mainstream labor movement could not imagine this. Therefore they focused their organizing on white workers and never strategically focused on Black workers. It was a famous failure with consequences that are obvious today.

So our one-line answer to the question "Why are radicals necessary?" is that the dialectic never quits. The dialectic that produces any social struggle, including ours, is between the objective conditions of oppression and exploitation on the one hand and the subjectively necessary presence of visionary leadership on the other. When combined in dialectical tension, they have the potential to produce a victorious movement.

There is no other way, in ordinary conversational English, to express these ideas. Furthermore, it is probably not an accident that most of us never received training in school for manipulating these essential ideas about how social change happens. Instead, this past generation of contingent faculty organizing has been heavily populated and often led by people who were educated in the organizations of the civil rights movement, the Black liberation movement, the socialist and communist cores of the anti-war movement and the student movement on campuses and likewise the organized groups of the feminist, LGBT, and women's liberation movements and the subsequent movements that emerged among Puerto Ricans, Chicanos, Native Americans and Asian Americans. Participation in these groups gave people not only some organizational tools and a view of themselves as organizers, propagandists, and actors on the stage of history, but also consolidated them in their radical critique of the world and forced them to revise their concepts and their practice, bouncing ideas about what works and what doesn't in regular discussions with others.

THE DOWNSIDE

On the other hand, if the minority of radicals only spend their time talking to each other and do not establish continuing ties with the typical majority, then any struggle generated will be weak and vulnerable to isolation and repression. This is especially true if the radicals are divided into small organizations that spend more of their time fighting each other or the existing union leadership than they do building the struggle against the employer, which can be very damaging. In fact, this sort of behavior can destroy organizations. The history of the labor movement is, unfortunately, full of examples of both errors, often referred to as the errors of sectarianism and dogmatism. Both flow primarily from isolation from the majority.

Many of these bad habits could be summarized by saying, as Max Elbaum did in his book *Revolution in the Air* about the New Communist movement of the 1960s–80s, that people became convinced that it was more important to have the most detailed and correct "line" or description of reality than to have large numbers of people who were listening to you. In other words, they might be right, but they were locked in a closet, alone. Faculty are particularly prone to committing this error, since we often find it easier to write, than to start an organizing conversation with someone we don't know.

Errors like these make relationship-building difficult. Not only radicals and leftists make them, of course. They matter more to us, however, because we lack the material resources of the corporate Democrats and the right wing. All we have is our relationships to other people, and anything that damages those relationships is crucial to us. On the other hand, the interaction between radical and leftist organizers and other less political activists can educate and train these less experienced folks both in techniques of organizing and also in how our struggle reaches beyond the immediate niche in which we conduct our tactical fights.

KNOWING ENOUGH TO LOOK FOR ALTERNATIVES
OUTSIDE THE IMMEDIATE SITUATION

John Hess often pointed out that a crucial moment in the development of the Lecturers' struggle was when they came into contact with the contingent faculty movement in the community colleges through the California Part-Time Faculty Association (CPFA). This organization brought together community college contingent faculty from across California. In the late 1990s, the CPFA itself had a broad enough vision to imagine the complete elimination of contingency, a status that was pervasive in the 100-plus campus system. They created actions like ACTION 2000 (A2K), a coordinated piece of guerrilla theater that mocked the then-common fear that the millennium would bring with it the collapse of computer networks (referred to as Y2K). ACTION 2000 was repeated up and down the state on behalf of community college contingents. It starred the "Freeway Flyer," a professor in academic robes with a chicken head and wings. The Freeway Flyer created a rowdy presence on dozens of campuses, culminating in a demonstration in Sacramento at the state capitol, where they dropped a Volkswagen on the capitol steps with this label on it: "Adjunct Faculty Office."

This openness to creative, militant, attention-getting tactics came partly from the radical political background of many of the leading activists. It motivated the CFA to create ongoing ties with the CPFA and with the Coalition of Contingent Academic Labor (COCAL) and ultimately with the Ruckus Society, all of whom at different times conducted educational sessions with the Lecturers' Council on topics ranging from direct action to the place of contingent faculty in the labor movement.

LEFTISTS AND DIRECT ACTION

Having radical ideas is likely to push someone in the direction of learning how to do direct action, such as A2K. Thus an experienced leftist is likely to know how to organize actions that will involve, engage, dramatize and publicize what workers want and ultimately challenge the powers that need to be moved. Such actions always challenge the status quo, patterns of customary behavior, and often the law itself. They are profoundly democratic, in the sense of requiring both participants and bystanders to make choices on the basis of their own values as the action proceeds. They exist outside the formal representational processes of bargaining, grievance handling and voting. Direct action can involve street demonstrations, guerrilla theater, or actions that members take together on the job, which can range from wearing T-shirts and buttons to selective job actions such as refusing to do free extra work or holding grade strikes. The ultimate direct action is the strike, the complete withdrawal of services. The most radical version of the strike is occupation of the workplace.

One important characteristic of direct action is the way it supplements official legally recognized representation. It is especially important and useful in contexts where representation has been heavily legalized and bureaucratized and where the subjects are usually submerged and invisible. In the heavily legalized context of sexuality, for example, ActUp demonstrations, PRIDE marches, women's marches, #MeToo, TimesUp, and "slut marches" confront the status quo with an alternative vision. Die-ins stop traffic and confront climate change denial. Direct action by contingents provides an opportunity to demonstrate to ourselves, to the union movement, and to the rest of the higher education community and society that we exist, that we have the capacity to speak in our own voice, and that we are the majority.

REPRESSION AND RESPONSE

We are not the first writers to recognize that radicals were often the essential ingredient to starting to get a group of workers in motion. John L. Lewis, president of the United Mine Workers of America (UMWA) and founder of the CIO, hired ("absorbed," is what some people call it) hundreds of Communist organizers to get the CIO off the ground in the 1930s. Since he had been a famous anti-Communist, someone reportedly asked him why a third of his organizers were known Communists. He is said to have replied, "Who gets the bird, the hunter or the dog?" The more conservative business unionist-oriented wing of the labor movement has actually created advice and teaching materials that explicitly tell organizers and union leaders that you have to find the radicals to get started, but then you have to figure out how to get rid of them (or at least marginalize them) once the union is established and you want a stable relationship with the employer.[3] This sometimes takes the form of suggesting that a radical can be allowed to manage the local's newsletter but should not be allowed to hold an executive office, or can be assigned to be an organizer—or even a salt[4]—but never regular staff.

Historically, this repression inside the labor movement has gone so far as to expel entire left-led local unions and national unions. An example is the American Federation of Teachers expelling locals—New York, Philadelphia and Los Angeles at slightly different times—that comprised a third of the total membership because they persisted in electing communists as leaders. This was on the initiative of American Federation of Labor President William Green, who was pressuring the AFT to "clean out" the reds. Various union constitutions also adopted provisions that banned communists. The Taft-Hartley Act of 1947 required union leaders to take a loyalty oath and swear that they were not members of the Communist Party. This ban was only removed from the AFL-CIO constitution in the last twenty years.

Given this history, which extends much further back in time, it is perhaps understandable that the vast majority of radicals, especially revolutionary socialists and communists in the labor movement, have been very careful about how open they are about their politics and organizational affiliations. In the case of the main radical organization of the 1930s–50s—the Community Party USA—this led to the vast majority of its activists in the union movement being secret, nonpublic members. During the Cold War, this allowed professional anti-communists and union haters to denounce them to the rank-and-file workers: "See, these people are actually criminal conspirators! You did not know they were communists. They hid that from you, and that shows they can't be trusted. They are really not fighting for the welfare of American working people." They were accused of being in the service of a foreign power, either the Soviet Union or the People's Republic of China, who were the Cold War-defined enemies of the US government.

This attack was effective, since millions of Americans had worked side by side with communists in the 1930s and 1940s, but had been unaware that they were Party members. This secrecy undermined the trust that thousands of these dedicated radicals had built up over the years through serious and effective leadership in the struggle against the employers in the short run and the ruling class generally. This is not to say that it's possible, or advisable, for radicals to be completely open all the time, in all situations, to everybody. Nor is it reasonable to tell people that the employer always deserves our full and complete revelatory honesty on all topics. Down that path lies suicide.

However, we think that history shows that it is better to err in the direction of openness to our friends, colleagues and fellow workers, because, over time, that honesty helps develop the real trust in our collective wisdom and in democracy itself, trust that is essential if we are to succeed in our struggle.

SCHOOLS FOR LEARNING HOW TO BE A LEFTIST

Someone who has put in time studying labor history is likely to take a long view of defeats. Much as we like to celebrate our victories, we know that they are all only temporary. Sooner or later, the wheel will turn and new conditions will arise. In the meantime, the struggle must continue. Something that leftists learn and can teach is how to look beyond the immediate win or loss and find the lessons that will provide the basis for the next strategy.

Sometimes radical ideas are injected into our struggle through contact with left-wing teachers in labor education programs. These are typically established at big public universities around the country as outreach to the labor movement and working people, just as the agricultural extension programs are outreach to farmers (and increasingly, agribusiness). A good book about labor education especially for women is *Cracking Labour's Glass Ceiling: Transforming Lives through Women's Union Education*, by Hanson and Paavo.[5] Though it focuses on Canada, it has two chapters on US women's labor education. The two main authors held labor educator positions at the University of Illinois. In classes at these programs, workers from all different kinds of workplaces, unionized or not, meet for extensive discussions of common problems and resources. Contingent faculty and graduate students who are considering unionizing share classes with nurses, bus drivers, custodians, electricians and other types of workers, and discover that labor has a common curriculum for all wage workers. Nor is labor education just one course; it is interdisciplinary, spanning the humanities and social sciences, with the difference being that it is applied. Not surprisingly, labor education programs have been continuously attacked over the last thirty years as the political context has moved to the right.[6]

Historically, labor and worker education has been conducted by independent labor schools like Brookwood Labor College in Katonah, NY and Highlander Research and Education Center in Monteagle, Tennessee; unions like the IWW, garment workers, the AFL and the CIO themselves through education departments; political organizations like the Communist and Socialist parties and their labor schools, and colleges and universities dating back to the Bryn Mawr women's school in the 1920s in Pennsylvania and the land grant university labor education programs after World War II.

In addition, there are independent labor education classes. John Hess worked with the East Bay Socialist School, an independent institution of the left in the East Bay, Oakland, in the 1970s and 1980s. At the present time, the Democratic Socialists of America are putting on well-attended socialist night schools. Both of these have predecessors dating back into the nineteenth century. Some organized left groups have sponsored initiatives that ultimately reach well beyond their limited direct contacts. Since the 1970s, the Solidarity organization has provided leadership for the continuing newspaper, *Labor Notes*, as well as biannual conferences that now attract thousands of participants from across the labor movement, and regional periodic educational conferences called "Troublemakers Schools." The Vermont Workers Center also holds Solidarity Schools and the East Side Freedom Library in St. Paul, Minnesota, is hosting the New Brookwood Labor School.

This practice of having a safe place where people can together explore ideas that are discouraged, ignored, and even forbidden in our daily contexts of work is essential to the future of our movement.

17
How Do We Deal With Union Politics?

Union politics is what most people are referring to when they say they don't want to get involved in their union "because of politics." Sometimes they mean "politics" in the sense of electoral politics which they don't agree with, but usually they mean the debate and discussion process and conflicts of personalities and loyalties that go along with decision making that affects them directly.

PERSONALIST VERSUS PRINCIPLED

In a union, there are really two categories of "politics." What is usually going on in *union* politics is that personal interests have become separated from collective strategies. Histories accrue rapidly: blame and resentment add to the heat, and wounds are inflicted. Issues of personal loyalty to leaders get conflated with issues of loyalty to the union. Accusations of anti-unionism get made on the basis of any criticism of the behavior of individual leaders. *Union* politics usually refers to some kind of terrible disappointment that has to do with these personal relationships.

It is true that the personal relationships people develop when working in a union context are powerful. When people work together both as professionals and in a union struggle, they will develop what are often the closest bonds they will have in their whole lives. The temptation is to compare this to a family. That is a dangerous comparison: a union is not a family. That is not its purpose, in spite of the fact that some unions have been led by family dynasties and some have produced family dramas worthy of Greek tragedy.

We need to be able to separate personalist politics from genuine principled differences. As contingents, we do have some principles that over much of our history, and across many unions, genuinely differ from the principles of some tenure-line faculty and traditional tenure-line faculty union leaders. These are real differences, and they are worth contesting.

Furthermore, racism gets loaded into the conflict whether or not there are non-white persons present in a group. Higher education has always had problems of race, class and gender at its core. Both higher education itself and higher-education unions in general are vulnerable to divisions over the question of racism because they are rarely honest about their own histories. Even when contingents are meeting among ourselves, it is an illusion to think we are all coming from an equal place and that we are all playing on a level playing field. One way to observe this is to note how rapidly trust disintegrates under pressure when some have more at stake, more to risk, than others. The cushion of resources, both economic and cultural, versus the responsibilities and demands that we face, are very different for men and women and for white people and people of color. The CFA has taken these issues seriously to the extent of developing an anti-racist and social justice initiative which includes representatives on each campus, a vice-presidential seat on the Board and a robust effort to incorporate these concerns into all aspects of the union's work. This effort is led in large part by contingents.

Racism can also be used as a strategic weapon by management to weaken a union. We discussed the way racism and the threat of unionism combined to guide middle management toward casualization of the faculty in Chapter 5. All other expressions of diversity—gender, age, ability, discipline, political perspective, nationality, immigrant status, habits of behavior—also feed the friction referenced by "union politics," and the result can be a disaster for collective goals.

SEVEN REAL CONTRADICTIONS

What follows is a list of seven real contradictions that represent actual differences over what a union should be and do. These are issues of principled politics rather than personalist politics. These issues can get confused and compromised by the overlay of personalist politics and feelings of being disrespected. Note that this list is not exactly the same as the list in Chapters 6 and 7, "What do we fight for?" which is a list of goals. This new list is a list of real material contradictions that face the union as it decides on strategy. Tactical choices around these issues can split a local union but have to be resolved to the benefit of a stronger union that can fight for *all* its members.

We will begin with trying to describe exactly what the contradiction consists of in each case. What issues are likely to split the union? We will follow this with a suggestion of what best serves contingents. We will use examples that some of our readers will recognize because we are trying here to channel the historical collective knowledge that the contingent movement has come up with so far. This is a work in progress, and we invite all our readers to join the discussion going forward.

Contradiction #1: What is bargainable?

The most obvious contradiction is about creating the bargaining agenda: what is bargainable and what is not? There are different conceptions of what is bargainable. How much energy can the union afford to spend to try to broaden the bargaining agenda?

In the years after World War II, a pattern of industrial relations arose during the Cold War that cast the union as a relatively secure but junior bargaining partner to management within a legally and customarily con-strained range. That meant unions gave up bargaining over anything other than the immediate wages, hours and working conditions of anyone but their members, and conceded to their employer, under "management rights," all questions of what is produced, how, for whom and for what price. This came to apply to services as well as manufacturing, and to the public as well as the private sector. This is sometimes referred to as "business unionism" or "service-model unionism," because this version of bargaining and then enforcing the contract was seen as the service that the union provided in exchange for member dues, like a business or an insurance agency.

The "junior partner" view of what the union can bargain is one side of the contradiction. The other side, never completely absent and rising now in the 2000s, is the view sometimes labeled "organizing model," "social movement unionism," or "bargaining for the common good." This seeks to organize and activate members to take an active role in bargaining and expand the scope of bargaining. It means bargaining things for members that are outside of their job (like housing) and also bargaining things that affect members of the working class, such as the quality of public services like education.

The history of our struggle as contingents shows that no matter what the law says or past precedent is, you can bargain anything that you have the power to bargain.[1] It is an issue of power, not law. The power in question is our power to force the employer to the table over anything that they don't want to bargain over. The only constraint is that it has to be something that your opponent across the table has the power to do. For example, your opponent may have the power to take a position on something (legislation, for example), even if they can't enforce it themselves.

Our movement history has shown that expanding the scope of bargaining has been good for subaltern groups (like us). It also helps us make stronger alliances with other workers and working-class people generally, as part of bargaining for the common good.

Contradiction #2: Negotiating pay, benefits, and compensation generally

One side of this contradiction wants to build the maximum amount of compensation onto whatever presently exists as a result of past bargaining, without primary concern for whom within the union gets it. The traditional goal is to get as much on the labor side—the biggest piece of the pie—as possible, through some degree of trade-off between pay and benefits. This goal takes up all the space on the union side of the table.

The other side of the contradiction has the goal that fighting for equality within the bargaining unit and raising up the bottom is more important than the total amount gained at any moment. This is because doing so increases unity in the union by bringing people closer together economically, and it undercuts management's continuing attempt to divide the union using, in this case, bargaining over economic matters.

If our goal is equality, the question for contingents is this: how do we get the union to bargain for more equal pay and benefits? This can be accomplished by getting contingent pay linked in some way to the tenure-line pay scale and then bargain the pro rata percentage up over time. Another way is to bargain lump sum raises. Percentage across-the-board pay raises are not a good way to get to equal pay. In fact, they increase the split between the bottom and the top.

Contradiction #3: Job security

The contradictions inherent in the question of job security are more complicated. Traditionally, unions bargained straight seniority systems as a way of restricting management favoritism. The problem arises as to who has access to the primary seniority system—whether it is Black workers, or women excluded from particular jobs or departments in a manufacturing plant, or the creation of second-tier faculty who are excluded from the tenure system. The contradiction is between those who see themselves as bargaining to defend the existing seniority system, exclusivist as it is, as a defense against employer discrimination and whimsicality, and those who see the primary fight as expanding and improving that system, which may include changing it to encompass the entire workforce, even those who have been excluded from it historically.

One obvious way that management attempts to weaken all job security systems is to introduce merit pay or pay for performance and, in the case of teachers, pay based on student test scores or complaints. Here the issue of job security overlaps with pay and benefit issues and has the capacity to throw faculties into hostile competition with each other.

We should fight to make the union prioritize giving current contingents preferential access to tenure-line jobs on a seniority basis and to increase the number of tenure-line jobs if contingents have reasonable preference for those jobs. Second, within our current contingent workforce, we should fight for seniority systems, but our goal should be to protect existing contingents in their current and past assignments to protect against favoritism among contingents by management and from hiring new people off the street as long as there are contingents available, qualified, and willing to do the work.

Contradiction #4: Other potentially splitting matters

Many other issues arise in bargaining that basically revolve around attempts by management to exploit potential differences within the union, especially in a combined tenure-line/contingent unit. These often have to do with issues of assignment, classification, load, evaluation processes, and other things that are of crucial importance to contingents. These can include offers to create new tiers such as full-time temporary or teaching-only positions for a limited number. They can also include limitations on tenure-line faculty overloads, either during the year or summer schools or intersessions. These may be presented as concessions by management at the table, but these partial "concessions" may not actually be good for the union or for us.

The general contradiction here is between those who see stability, maintaining the status quo, and minimizing irritation to lower-level administration, department heads and senior tenured faculty as being the priority, and those who see the proper goal of the union to be elevating conditions and rights so that those at the bottom achieve greater equality and unity and participation in the union.

As before, we would argue that the aggregated historical wisdom of the movement has shown that contingents are better off if the union fights for improving the rights of all contingents instead of being satisfied by getting small individual opportunities for advancement for particular contingents, especially if they are going to be chosen by the administration. We fight for a rights-based, not a privilege/opportunity-based regime. The same principle could apply to our response to other partial concessions that management might make in bargaining. Likewise, that principle applied to the question of overload assignments would lead to fighting for minimizing tenure-line faculty overloads, which not only degrade the capacity of tenure-line faculty to do a good job on their existing full load, but also take classes from contingents. It would also apply to issues of evaluation: even though evaluating contingents appropriately (meaning, not just with student evaluations or self-study) is a large administrative task and an irritation to those faculty assigned to do it. The principle we should fight for is raising up the bottom toward equality, which would mean the right of all contingents to be regularly evaluated in the same way and on the same basis as tenure-line faculty, along with the possibility of contingents being chosen and paid to do some of that evaluation.

Contradiction #5: How the union structures itself

In general, all union structures deal in some way with the two contradictory demands placed upon unions. One is to provide an ongoing institutionalized vehicle to protect and defend the members on the job over time. The other is the need for the union to always be the embodiment of the social movement of the workers fighting for better conditions on the job, in life, and as part of the broader working class in the whole country. In a word, this is a contradiction between the need for an institution on the one hand, and the need for a movement that involves the members as part of the working class.

Most of what is called representation would fall on the institutional side of the union's structure. But representation also includes leading and mobilizing members to fight in a collective way on our own behalf. This is true even in grievance situations, using tactics like mass petitions, quickie work stoppages, etc. These are also instruments of representation that have more of a movement character. This contradiction is built into the structure of a union. This will show up in its constitution and by-laws; how it manages meetings, committees, communications and elections; educates its members, celebrates its wins (or criticizes and examines its losses) and its history; how it organizes new members; builds or fails to build a steward system, etc.

What should be our attitude toward union structures? As contingents—a majority second-tier workforce both at our job site and within our unions—our interest is in broadly democratizing unions and generating the maximum feasible participation, especially among ourselves. That means equal access to all elected offices and maybe affirmative action-reserved seats for contingents on governing and decision-making bodies. Since our active participation in the unions is more difficult—timewise and financially, and because it involves political risk given that we do not have job security—we need conditions that facilitate our greater participation. This means stipends, for example, and political protection from retaliation. The union also needs to invest financially in us in terms of sending us to trainings, conventions and conferences, and to provide support to organize ourselves within the union—because even within the union, contingents often feel constrained to speak their minds fully in front of tenure-line faculty (and especially in unions that may include department heads).

This last—department chairs in the union—is a major factor in both dividing the union and minimizing free, open participation in those unions in which department chairs have full rights. These people also tend to rise into union office because of their experience and respect, especially in places where they are elected. It can and does happen that a tenure-line faculty member becomes a department chair but remains an activist in the union—for example, a grievance committee member—and then has to carry a grievance on behalf of a contingent member or may even be the person against whom the grievance has been brought. If contingents are to have full democratic rights, supervisors who are bargaining unit members must be restricted from having leadership roles in the union.

A number of other contradictions arise from the structure of the union. Contingents, organized among ourselves, need to discuss and decide how to deal with them collectively. One is whether people should run for office, individually, or as part of a slate linked to pro-contingent tenure-line faculty, or as a contingent slate. This question also comes up when the union talks about hiring staff. Who writes the job description? How is the pool of possible hires gathered? Are they outsiders or are they members of the bargaining unit? Who does the actual hiring?

Then there is how we behave tactically in meetings and other gatherings and how much and with what tone we press our agenda, how many of the microaggressions that we face in those contexts we ignore, and how we pick our shots on what to fight about.

Here is John, blowing off steam to the then UPC President Stuart Long, who came to a meeting of the official statewide Lecturers' group. He had previously agreed to leave if asked so that people could discuss freely. John, sensing that this was a moment when a "safe space" for discussion was needed, did ask him to leave. Long responded defensively and refused. The president was evidently clueless about the very people whom he was supposed to represent. John writes:

> You agreed that if an issue arises that we are [*sic*] uncomfortable about your presence, you would leave until we invited you back. I asked you to leave because I felt we could have a much better discussion of [the bargaining team] without you. I knew that if you stayed, all the anger and alienation from full-time people that Lecturers feel would come out and muddy the waters ... For Lecturers, all the indignation and alienation of being a Lecturer comes at the hands of the full-time faculty, not the administration. It's a great leap of theoretical understanding to see that it really comes from the administration, that full-time people are also caught up in a system that humiliates them. My immediate destiny is in the hands of two or three people in my department, so every time I see them I am in the presence of a nearly absolute power exercised over my life at San Francisco State. Every look, anything they say to me or about me, or anything they say about my competitors has to be evaluated. It takes a great psychological toll ... Among ourselves, we could have had a much more intelligent, rational discussion. [Letter from John Hess to Stuart Long, October 20, 1981]

Contradiction #6: How the union presents itself to our allies and the public

Traditionally, higher-education faculty unions have gone out of their way to present their demands as student-centered, assuming that most people see college professors as an already privileged group and that to present our demands as workers would alienate the majority of the public, even the working-class public. In other words, traditionally, we as faculty unions have presented ourselves and our demands not as fellow workers to a working-class public, but as "professors," professional academic service providers who are ourselves outside the working class and deserving of special deference—academic freedom and tenure—because of our special role with regard to students and research. The special academic service that we provide is the basis of our appeal for support. But the reality is that today, most academics are contingent workers who should not be embarrassed to put forward our own material needs as legitimate and as part of the working class, as well as linking our welfare to the welfare of the students we teach, since "our working conditions are their learning conditions," as we have said before. The contradiction is between professional elite service and working-class solidarity: Which do we seek to build?

The contingent faculty movement has just begun to push actively for this revision in how faculty unions should present ourselves and our demands. A resolution of this contradiction in the direction that we suggest would have an impact on our relations with other unions, other contingent workers and their organizations, and the media. We would do well to follow the example of some of the best K–12 unions like the CTU in Chicago who have moved in the direction of presenting themselves as fellow workers in the service of a working-class public.

The Program for Change that has been circulated in the contingent faculty movement lays out steps towards greater equality and the end of the two-tier system. It is basically a long-term bargaining plan. As an intervention into the contingent faculty discourse, it represents a path to the most complete resolution of the above contradictions that we are aware of. However, as we noted, it comes out of Vancouver, in British Columbia, where the conditions for bargaining are more favorable.

Contradiction #7: Contingent-only unions and bargaining units

The contradictions surrounding contingent-only unions or contingent-only bargaining units within larger union locals have some differences but more similarities to those in joint units. The differences, obviously, are that contingent-only bargaining units bargain only for contingents. However, they remain influenced by the terrain of the existence of tenure-line faculty, who may or may not be unionized at their institution. Thus, in shadow form, many of the issues mentioned above arise for them as well, especially in terms of the management attempting to use tenure-line faculty interests, as they characterize them, against contingent faculty at the bargaining table. This is not a trivial difference, but at the same time it is not a fundamental one, since everywhere we occupy a joint workforce and face the same strategic opponents regardless of how we are organized or not.

It is important to make clear that while the majority of unionized faculty in the US are in joint tenure-line/contingent units, this is because they are in the relatively few states that have adjudicated joint units as the default (California, Washington State, Oregon, New York) and these states also have the most organized faculty. However, the majority of actual bargaining units in the US are split into contingent-only or tenure/full-time line only, and the growth nationally has been largely in split units, especially in the private sector.

Some of these split units even have different national unions representing the two or more units at a single campus. The contradictions discussed in this section are not unique to joint units. They will arise in some way, at some level, in contingent-only units as well.

Finally, some of the contingent-only units face the problems mentioned above at the level of their state and national affiliates, as well as at the local union level.

A LAST WORD ON UNION POLITICS

We may seem to have strayed far from what the average person would describe as "union politics." As we noted at the beginning of this chapter, we do not want to either deny or avoid the fact that the experience of participation in the pressure chamber of union activity, which is above all a fight and deserves to be respected as a fight, can become loaded, poisoned and corrupted by personal hurts. However, our purpose is to draw attention away from the personal aspects of the big fight to focus on the deep principled differences that underlie the contradictions inherent in union representation. Focusing on these principled contradictions, and strategizing to address them through collective positive action, can in many cases reduce or deflect the risk of personal wounds. The central contradiction, of course, for us, is power in the union: who has it, who does not, and why.

PART VI

USING THE POWER WE HAVE

18
Hopes and Dangers

Coming out of the Covid-19 pandemic in 2021, we can expect to see greater extremes of inequality. Setting this in the context of the global climate crisis, we predict that the struggle over the role and character of higher education will be intensified. But a living planet needs a sustainable means for producing and passing on knowledge. "Sustainable" means sustainable institutions, providing jobs that can make life sustainable for the people who do the work. This is the challenge that our movement confronts.

To that end, we have divided what we see in the current landscape into signs of hope and dangers for the movement. One the one hand, we have a rising public consciousness of the inequalities that have steadily sharpened in the last thirty years. Examples of this growing awareness include Occupy, the Black Lives Matter movement against systemic racism and police and criminal justice abuse, the #MeToo movements against sexual violence and rape culture, the Bernie Sanders presidential campaigns which were largely directed against inequality, the teachers' strike waves of 2018 and 2019, the voter registration drives of 2016, 2020 and the Georgia Senate run-off in 2021, and the rise of the Democratic Socialists of America, which has normalized the word "socialism" in the United States for the first time since the Cold War. On the other hand, we have the white supremacist neo-fascist movement that found its leader, protector and mouthpiece in Donald Trump, the 70-plus million voters who supported him and his party in the 2020 election and the Republican majority in Congress who failed to indict him after he incited the January 6, 2021 attack on the Capitol. When a dialectic is as polarized as this, the famous quote that applies is from Gramsci: "The crisis consists precisely in the fact that the old is dying and the new cannot be born; in this interregnum a great variety of morbid symptoms appear."[1]

SIGNS OF HOPE

1. There has been a rise in contingent faculty organizing. Thousands of contingent faculty have been organized into unions in the last few years.

2. This increase in organizing has come primarily from the grass-roots, not from the top down. Organizing activity builds on itself. Typically, contingents self-organize and then seek a union to represent them. The current exception to this is the SEIU national campaign.

3. Two distinct strategies have emerged from this organizing. One is the Metro strategy, reflecting the geography of the workforce and giving people contacts across institutional and union lines. The other is the Inside/Outside strategy, a strategy that people come up with when they have to organize themselves. This strategy creates leverage within larger organizations. Both of these strategies focus away from the bargaining table and involve more than one union, institution, or outside organization.

4. New leaders have emerged who are younger, often women and/or people of color, reflecting the actual workforce more accurately than leaders have in the past. Many of these leaders come out of graduate student organizing. New leaders bring new tools, especially social media, which produced viral discussions of the conditions of contingent faculty (for example, the meme of "The Homeless Adjunct"[2]) and created National Adjunct Walkout Day in 2013.

5. Contingent faculty are now a much higher priority for national organizations. There are formal committees and caucuses within many of the major unions and disciplinary organizations.

6. These committees and caucuses have taken on independent outside-facing roles such as participating in COCALs, NFM, other groups such as the Coalition on the Academic Workforce (CAW) and the Campaign for the Future of Higher Education (CFHE) and coalitions formed around unemployment insurance.

7. Since the 2018–19 Arizona, West Virginia and Oklahoma et al teacher strikes, there has been a continuing upsurge of organizing and activism among our brothers and sisters in K–12, including Chicago, Los Angeles, Denver, Oakland and Puerto Rico. This upsurge arose around issues of bargaining for the common good, and it shares with our struggle the basic idea that teachers are part of the working class and our teaching conditions are our student's learning conditions.

9. There is a general backlash against neo-liberalism in education, from charter schools[3] to for-profits and the run-it-like-a-business model. It was an issue in the 2020 US presidential elections and continues to be an issue in the Biden administration with demands for "Free College" and debt forgiveness. There is criticism of privatization generally and hostility to the rise of gig work specifically.

10. The pandemic has changed the terrain and revealed how some of our long-standing demands as precarious workers apply to all essential workers and the working class as a whole. Examples of these are unemployment insurance, Medicare-For-All, and paid maternity, family and sick leave. In particular, our need to be able to exercise academic freedom is like any worker's need to be able to exercise free speech on the job without fear of retaliation. These demands are universal and should apply to any worker in a sustainable society.

SIGNS OF DANGER

Many dangers can also be understood as signs of hope. Some are, in fact, good dangers to have. Nevertheless:

1. A lot of the organizing that is taking place is exhibiting a lack of respect for democratic process, for mass education, and for democratic functioning. It is too narrow and often dominated by non-contingent faculty, staff organizers and union officers.

2. In our list of hopes, we mention two strategies that are in operation now. We have criticisms of both. In many places that are organizing under the Metro strategy, the Metro organizations are letterhead only or functionally nonexistent. They are not really owned or run by contingent faculty themselves. The Inside/Outside strategy runs different risks. While the Metro strategy implies a confrontation with higher education as a whole, it does not confront power directly. The Inside/Outside strategy confronts power directly because it emerges as a result of having to self-organize separately within the overarching organization in order to have leverage. The risks here are that organizing can be completely suppressed or, on the other hand, it can be co-opted and give up its capacity to speak as an independent collective voice within the overarching organization.

3. There is often an over-focus on measurable benchmarks and legal pro-cesses of organizing and representation: getting cards signed, recognition, getting first contracts, collecting dues. This over-focus, though understand-able, often short-circuits effective movement building: deep education, one-on-one conversations and leadership development among the rank and file themselves. This has resulted in losing elections, pulling campaigns, or failing to get a decent first contract or even a first contract at all.

4. We still don't have a national legislative strategy in practice. We have lists or goals from the SEIU, NEA, AFT, AAUP and others. They have each sep-arately acted on pieces of a national legislative strategy, but they have not contributed significantly or sufficiently either to a joint effort or an effort by an outside organization. While the NFM was able to mobilize the major unions to support us on the issue of unemployment benefits, that support was not enough to win more than a partial victory.

5. Consistently, the lack of resources has limited the capacity of the outside part of the Inside/Outside strategy. None of the unions, much less the dis-ciplinary organizations like the Modern Language Association (MLA) and the Organization of American Historians (OAH), will prioritize contingent issues sufficiently in their budgets without pressure. This pressure cannot come just from inside; has to come from outside as well. Only a stronger, more consistent and better resourced outside group, whether it is the New Faculty Majority (NFM) or successor organizations, can bring that pressure to bear.

6. Different viewpoints exist generationally. Today, there is a need to pass the baton. Younger contingents are now part of a different context from those who began the contingent faculty movement back in the 1970s. The danger is that this transition will not take place in time.

7. Another danger is that the existing organizations and institutions that the movement has built—caucuses, committees, etc.—are still dispropor-tionately led by older white male activists and need to be more proactive in passing on the leadership and mentoring the next generation of leaders than has often been the case. Many of these younger activists are women, LGBTQ and/or people of color.

8. There are also divisions between people who work in different subsectors of higher education: community colleges, private liberal arts colleges, uni-

versities both public and private, and adult education. The prestige—or lack thereof—of an institution can be leveraged by management to divide people who should be engaged in a common effort.

These two lists demonstrate the real existence of a movement, with some achievements and some bumps in the road ahead. On the one hand, we know that casualization must be addressed as a totality, not as an array of separate solutions to management problems that can be reverse-engineered. This means that we have to address the system as a whole. On the other hand, we recognize that our demands are universal, which means placing ourselves deeper into the labor movement, where other workers and workers' organizations are already struggling to win them. We have two strategies at work (the Metro strategy and the Inside/Outside strategy), which are effective but limited in scope. We have no clear national (or international) strategy to work with. This broadens and deepens our debate and pushes us toward our Blue Sky agenda.

A SIMPLE ABSTRACT IDEA

What our movement has moved away from, correctly, is a focus on various versions of the victim narrative and arguments against contingency on the basis of morality, justice, logic, or common sense. Despite being presented in books, articles, poetry, reports, podcasts, testimony and movies, those arguments have not won the day. This is not to say they have failed, since each of these brought more people into sympathy with the movement. However, they have not stopped the increase in casualization. They did not unlock the combination of incentives and fears that made casualization a solution for management faced with budget cuts, flexibility demands, the threat of unionization, and hiring from a labor pool that looked more and more like the working class: older, more women, more people of color, and, at least at first, more willing to settle for less in the hope that things would improve.

One reason why they have failed is that they are appeals to power that is based somewhere else: legislatures, administrations, the public and to some extent our students. This is power we can influence but not control. These arguments are not strategies, plans about how to do something. They are attempts to get someone else to do something.

What these arguments do show, however, is that there is uncertainty among leaders in our industry. Consultants, researchers, policy makers, professors of education in the US and around the world are responding to the crisis by repeating these same arguments, all noting that contingency and the majority contingent faculty is a problem—not just for us, but for them. They recognize it and analyze it, but it is not really in their power to do anything about it.

The people who have the power is us—ourselves. The arguments that can succeed are the ones directed at our power and our ability to build it. Like any workforce, our power comes from our work and our ability to do it or not. Our work is essential work; we are the majority of people who do it; we can choose to do it or not. Therefore we anticipate arguments that are actually strategic organizing discussions provoked by troublesome questions, such as the questions we have identified in Chapters 10–17 and the many others that will always come up. These will be arguments about strategy, about what to do and how to do it.

We have talked about two strategies that are shaping our movement today. One is the Metro strategy, an approach to organizing a whole workforce across institutions and unions. The other is the Inside/Outside strategy, an approach to self-organizing groups of contingents who are represented by some larger body—a union, a disciplinary organization—in order to speak with one voice to both that organization and the public. Both of these are part of our movement but they are not themselves a strategy for the whole movement.

A strategy for the whole movement will have to rise out of the transition that is taking place around us right now, in a world where change has been accelerated by the pandemic. The scope of our next strategy will probably be positioned between the Metro and I/Os strategies on the one hand, and our Blue-Sky goals on the other.

The Blue Sky goals that we describe in Chapter 6 will not change; they will continue to be what we need in order to do our work the right way. What will have changed is the reality in which those goals will get actualized.

Examples of aspects of an intermediate strategy would include putting more of our effort into local labor movements, recognizing that our demands are universal demands of all working people. This includes supporting labor struggles of other workforces. Every campus has multiple workforces, many of them unionized; a campus labor coalition would be a low-key place to start. An intermediate strategy would also include more focus on higher education internationally. Education has, in recent years, become globalized in terms of research, faculty and student bodies, and exhibits a wider range of conditions for faculty, good and bad, than are found in the US. Internally, we need to connect with the 70–80 percent of contingent faculty in the US who do not yet have a union and plan how we are going to reach them, both through our power in the national unions and through independent action. All three of these are ways to contribute our power and draw on the power of our allies.

The institutions of higher education in which we work may look very different in the next few years. They will probably be more globalized, more online. There will be major battles about funding and curriculum. What we can be sure of is that most institutions will try to eliminate tenure and universalize contingency. That will be our big battle. The piece we can control is our own power, which we create through organizing and then deploy when it's time to make our move. The purpose of this book is to contribute to that power.

Essential Terms

Academic freedom. The classic AAUP definition is found in the 1940 Statement at aaup.org. In common usage, *academic freedom* has been used to mean a right to the full freedom to research, publish and teach in one's field without fear of intimidation or retaliation. It also had been defined to include the rights of academics to engage in free speech outside their institution. The true edges of academic freedom remain in dispute, both in common usage and in the courts and legislatures. For us, academic freedom for academics also relates to the academic freedom of students to hear divergent views, learn, discuss and act on them.

Administrators, managers, supervisors, employers, bosses. These are all terms for individuals in position to exercise authority over us as contingent faculty. This would be contrasted with colleagues, fellow faculty, etc. In some contexts, the term "administrator" also applies to people who do administrative work without any supervisory or managerial authority.

Bargaining unit. The adjudicated grouping of employees who are covered by a particular collective bargaining agreement negotiated by a particular bargaining agent (union).

Bargaining election. The election that affirms or denies the will of the majority of a workforce to choose to bargain collectively or/and what bargaining agent (union) will represent them. Legally, it is conducted by a government agency or informally conducted by another neutral party. Sometimes called a "union election."

Bargaining for the common good. Bargaining or otherwise fighting for issues that are beyond that which is most narrowly seen as wages, hours and working conditions, and which impact broader society. See www. bargainingforthecommongood.org

Benefits, "fringe" benefits. Any compensation other than direct salary and wages. Can include health insurance, retirement funds, paid leaves of any sort, sabbaticals and others.

Cervisi Decision. Full name: *Gisele R. CERVISI et al., Plaintiffs and Respondents, vs UNEMPLOYMENT INSURANCE APPEALS BOARD*. The California Court of Appeals decision of February 1, 1989 which declared that a teaching assignment *contingent* on enrollment, funding, program changes, or bumping by a full-time faculty member is not "reasonable assurance of re-employment" as required by the federal unemployment laws and therefore that instructors so situated are eligible for unemployment benefits as a group. This decision unfortunately remains unique in the US.

Class. As we use it in this book in the Marxist sense, *class* refers both to a person's structural relationship to the means of production (capitalist/owner versus wage earner/worker) as well as to the conscious self-reflective activity of an individual and or group identifying a part of a class. Thus, the expression, "a class *in* itself" (structural material position in the economy) and "a class *for* itself" (conscious activity of individuals and groups identifying as part of a class). In our case, the "means of production" is not a factory equipped with tools, but cultural and social production and reproduction in a system of higher education. In this book, a central continuing discussion is the issue of the consciousness of faculty. The vast majority of faculty are now structurally members of the working class. Nevertheless, many hold to the belief that they are members of an independent professional middle class but, as contingents, are degraded members of that class. The tension between the reality and the belief is a critical issue for organizing. This use of the term *class* is distinct from many other uses of the term *class* in popular culture which is often based on income, culture, education level, etc.

COCAL. Coalition of Contingent Academic Labor. When it is Chicago COCAL or Connecticut COCAL, using the name of a city or state, it refers to a coalition of groups or individuals representing contingent faculty in a specific region. A local COCAL is often the base for a Metro strategy. When it is COCAL International, it refers to the various groups, COCALs, unions and independent groups that come together in the biannual conferences involving contingents in the US, Canada and Mexico and the International Advisory Committee.

Consultative power/executive power. Consultative power is the mandatory right to consult and advise with an entity that has executive power or final authority to decide and implement.

Contingent. Without secure employment, used in higher education synonymously with precarious, adjunct, non-tenure line, and in the CSU system officially referred to as "Lecturer."

Collective bargaining rights. First, referring to the legal right of a category of employees to bargain collectively with their employer, and the duty of the employer to bargain with the elected bargaining agent. These rights apply to both a category of employees (like public-sector workers in a certain state) and also to the specific organization that gains those rights on behalf of a particular group of employees.

Decertify. The legally established process by which a majority of employees in a bargaining unit vote to cease having a particular union (bargaining agent) represent them in collective bargaining with the employer.

Dialectic. The concept, flowing from Hegel and Marx, that changes in ideas and in the material world take place as a result of the conflict (contradiction) between two opposing forces. Out of this conflict (contradiction) comes something that is neither the mere addition or subtraction from the original pair of forces, but rather something new in the world. This general theory of how change happens is referred to as "the dialectic" and it is asserted by many to be the appropriate term to describe transitions in both human and natural history.

Direct action, strike, job action. Actions taken collectively by workers away from the bargaining table (not bargaining or grievance negotiation actions) with the goal of pressing an employer or governmental body. Examples would be wearing a button, signing a petition, and full or partial work stoppages.

Dissertation. Final extended original research document to complete a terminal or doctoral degree.

Due process, just cause, cause. As originating in English common law and listed as a basic right in the US Constitution, means that a fair, impartial and non-discriminatory judicial process should be applied in all cases, publicly. The most important part of due process in employment and labor relations is the concept of *just cause* for discipline or discharge, often referred to merely as "cause." Due process rights are included in tenure but must be bargained in a contract for non-tenure-track contingent faculty.

Exploitation. In the Marxist tradition, exploitation means the taking of the value of the work produced by a worker by the owner, the capitalist employer, so that the worker receives less than the full value of what they produce. In common parlance, exploitation is often used as a synonym for any oppressive, unfair condition or even the "exploitation" of natural resources.

Exclusive representation. The characteristic of US employment relations law that assigns the right to collectively bargain for a particular group of employees (bargaining unit) solely to one organization (union). This is in contrast to employment relations regimes in other countries such as France.

Fair share, agency fee, agency shop. A version of legalized union organizational security that requires all employees in a particular bargaining unit who gain the benefits of a collective bargaining agreement to pay a fee to the bargaining agent (union) for those services that they receive in bargaining and grievance handling. This is distinct from membership in the union, which remains voluntary and also includes other services, political representation and the right to engage in decision making as a member.

Full-time, part-time. Terms used to denote the amount of a full load that a particular employee is classified as working for a particular employer. Some employers classify everything below 100 percent as part-time. Others define part-time as under 50 percent. The definitions are not consistent or stable. These terms do not always actually describe what people are doing, but rather their classification by their employer, since people can work more than full time for one or more employers in one or more positions. The terms are commonly in use and so cannot be avoided, but have limited usefulness, so we have avoided using them as much as possible. Full-time and part-time can be tenure line or not tenure line, secure or contingent, and that is true in the CSUs.

Gig economy. A twenty-first-century term used to mean unstable, contingent, often part-time employment, typically mediated by a computer platform.

Governance, institutional governance, shared governance. The term that is usually applied in higher education to decision-making power over various aspects of the institution. Historically, by custom in higher education in the twentieth and twenty-first centuries in the US and abroad, governance was supposed to be "shared" between faculty and those who were delegated to perform managerial and administrative duties. In the recent past, the issue of governance and shared governance has become a matter of lively dispute. The concept of shared governance has been explicitly challenged by casualization, the rise of the for-profits where there is no pretense of shared governance, and the overall erosion of shared governance. Like academic freedom, the "gold standard" of shared governance is found in the AAUP's *1940 Statement.*

Hegemony. The power of any ruling regime to control not just the material environment but also the ideological and cultural environment. Universities have hegemonic power, as do religions, states, etc. The hegemony of any regime is never fixed and permanent; it is a terrain of struggle. Antonio Gramsci is the main source of the modern concept of hegemony.

Inside/Outside strategy (I/Os). Any strategy for a less powerful group to gain power within a larger group of which it is a part, but which it does not control. Examples of this "larger group" would be a union or a political party. People in a subgroup have to self-organize separately in order to have power inside this overarching organization. If they are merely unorganized individuals within the overarching group, they will not be able to overcome their second-class status even if they are the majority. The success of the "inside" strategy is dependent upon the previous existence of the "outside" strategy. The concept has its roots in Gramsci's ideas of workers' councils, which organized to gain control over production. In our case, the contest is over the hegemonic role of the university in producing culture, learning and knowledge.

***Janus* Decision**. The 2018 US Supreme Court decision ruling that employees in the bargaining unit of a union in the public sector with exclusive bargaining rights could not be required to pay an agency fee for negotiation and representation services that the union is legally mandated to provide them. This decision was expected to greatly weaken public unions and lead to a massive drop in membership, but as of 2021 this has not occurred because unions have engaged in member education and internal organizing to counteract it.

Jim Crow. The legal and de facto practice of racial segregation in all public and most private spaces that arose after the defeat of Reconstruction in 1877 and was most formally ensconced in law by the state constitutions passed in most Southern states between 1890 and 1910 and was approved by the Supreme Court in *Plessy vs Ferguson*, "separate but equal." Jim Crow was not legally overturned until the civil rights movement of the 1950s–70s. These Jim Crow conditions were also applied throughout the Southwest to people of Mexican descent, and in some areas, to Asian and Indigenous people as well.

Labor movement. We use the term *labor movement* in the broad historic sense to include not just collective bargaining organizations ("unions" in the American definition) but also working-class organizations such as workers' centers, tenant unions, working-class political parties, cooperatives, etc. Historically, the term *labor movement* was used to refer to all of those, but in the context of the US in the twentieth century, the definition was narrowed to mean only the union movement and only that section of the union movement that engaged in collective bargaining under the law.

Layoff, termination, reduction in force (RIF). For contingent faculty, these all amount to the same thing. Other terms used for this are de-scheduling, loss of classes, bumped out. Any and all of these should make a contingent eligible for unemployment benefits as they are de facto evidence of lack of reasonable assurance for re-employment.

Lecturer. The term for most contingent faculty in the CSU system. When used in that way, it is capitalized because it is a rank, like Assistant Professor, etc.

Leftist. As used here, anyone whose vision of social change goes beyond liberal reforms but rather advocates radical changes in a progressive, democratic direction.

Lockout. A work stoppage initiated by the employer, as compared to a work stoppage initiated by the workers such a slowdown, strike, etc.

Manager. One of the terms we favor in this book for those with supervisory or managerial responsibilities, as compared to the commonly used term "administrator," is manager. The term "administrator" is commonly used but it is unclear because it includes workers who do not in fact have managerial or supervisory responsibilities, but who are simply carrying out necessary administrative work such as various forms of record keeping, and processing of communications. Some of them are unionized. We do this in order to highlight the class line within the institution, which is often blurred in higher education discourse.

Marxism, Marxist. From Karl Marx (1818–83), the philosopher and author of *Capital*, a study of how capitalism works. His name has come to be attached to a wide range of theoretical approaches to learning in fields as disparate as history, sociology, literature, evolutionary biology and economics. What these all have in common is an approach that is historically grounded and materialist. In the US especially, Marx's association with radical politics and communism made the label "Marxist" a red flag that made hiring, to say nothing of promotion, unlikely, for anyone who claimed it.

Metro strategy. Organizing a regional workforce (usually commute-distance, since contingents typically commute among a set of regional institutions) rather than on the basis of institution-by-institution organizing. The idea is to organize the workforce and negotiate standards for all members of that workforce. As of current writing, the SEIU has taken the lead in a Metro strategy approach. The AFT and USW have also initiated Metro strategy campaigns.

Negotiate (bargain). The process of two or more parties sitting at a table (meaning a neutral place where communication can take place, but often a real table) as at least temporary equals to determine a resolution to a dispute. It is assumed that each party is prepared to both demand and concede on particular points. The willingness to demand and concede is part of what is considered "good faith." Imposing the will of one party on the other party is not part of negotiation; it is order giving or *imposition*. Negotiation can be individual or collective.

Program for Change. A model and guide for bargaining developed by Jack Longmate, contingent activist from the State of Washington, and Frank Cosco, former president of the union at Vancouver Community College in British Columbia. The contract at VCC nearly eliminates contingency. The Program for Change illustrates how to incrementally bargain toward that goal. However, it was bargained under Canadian and BC legal and political conditions. Visit the website http://vccfa.ca and click on "Resources" (accessed April 30, 2021).

Proposition 13. The 1978 California proposition that capped property taxes for residential and commercial properties and started the steady decline of property tax revenue, which affected school funding. California went from near the top among states in spending per student to near the bottom. "Prop 13" also required a supermajority to approve taxes or the state budget.

Reconstruction (Radical). That period between the US Civil War (1861–65) until 1877, when a serious attempt to provide and protect full political, economic and social rights for freed people in the former slave states of the Confederacy. It included the election of Black men into government offices from the local to the US Senate. The 13th, 14th and 15th Amendments to the Constitution (known collectively as the Civil War Amendments) were passed in this period. Radical Reconstitution did not include the promised reparations ("forty acres and a mule") for formerly enslaved people. That, combined with the withdrawal of occupying Union troops as part of the settlement of 1876, sealed the end of this period and marked the beginning of the Jim Crow era and re-establishment of power to the Southern white plantation ruling class.

Representation. A group delegates (chooses) a representative (an individual or organization) to speak, advocate, or negotiate on their behalf. If the representative is legally recognized under labor law, then it, the individual or organization, has a duty to fairly represent all of the individuals and groups that have selected it. Thus, a workforce (bargaining unit) selects a bargaining agent (a union, and individuals in leadership or staff positions in that body). From this legal relationship flows the concept of "duty of fair representation." Legally, in labor relations and in this book, representation means the right and often duty to bargain or otherwise "represent" vis-à-vis the employer. This is distinct from the common usage of the term *represent* or *representation* which can include asserting something as true or to informally claim to speak on behalf of someone or something.

RIF. This stands for reduction in force, the management term for any action that lowers the headcount of employees, for whatever reason.

Salt (noun). A person delegated, and perhaps partially paid, by a union or political organization to go into a workplace as an employee with the explicit purpose of organizing or at least gaining information for organizing. Also used as a verb.

Scope of bargaining. This refers to the permitted range of issues that can be legally bargained under a particular bargaining law. At the national level, in the private sector, it is also a matter of continuing dispute in practice. The scope of bargaining is different in different states for public employees. Bargaining issues are divided into mandatory (they must be bargained), permissive (can be bargained if both sides agree to do it) and prohibited (things that cannot legally be bargained, even if both sides agree). Generally, unions have struggled to expand the scope of bargaining and employers have fought to narrow it as much as possible. Within the contingent faculty movement, there are examples of successful bargaining outside the technical scope of bargain when the union was strong enough to force the employer to bargaining over it.

Tenure. From Latin, *tenere*, to hold or possess. In academia, this applies to someone's presumed right to continue (hold on to) employment after a five- or six-year tenure-track probation period. This right of continued employment can only be abrogated on the basis of due process just cause, or layoff for financial exigency, or the closure of an entire program. In other words, tenure is the historic academic term for the right to due process just cause dismissal. It has also in academia been linked to a right to shared governance in an institution and academic freedom. The term is often linked to the promotion of Assistant Professor to Associate Professor rank. The state of being *without tenure* is what we refer to in this book as contingent or non-tenure line, the situation of the majority of college and university faculty.

John Hess: A Life in the Movement

Sometimes we ask ourselves if our contribution actually makes a difference. Does it matter if we make one more phone call, show up at a meeting, or help a colleague file a grievance? As individuals, our work is invisible in the crowd. We are nobody's heroes. But the consequences of our efforts make a vast, even life-or-death difference to thousands of people like us.

John's life can be told as the nearly typical story of a contemporary academic intellectual, navigating the changes of the last forty years, trying to make a living in an increasingly tough job market, all the while moving politically further and further to the left. He grew up in the religious Mennonite culture of eastern Pennsylvania, went to Lehigh University and then was drafted into the military in the early 1960s and found himself in Germany. The years in Europe, where he met and married a German woman, introduced him to socialism. He came back to enroll in graduate school at Indiana University where, his friend Chuck Kleinhans says, he "circulated in the fluid movement of New Left activism, radical political filmmaking and a new critical intellectual climate." John, Chuck, Julia Lesage and other friends started the journal *Jump Cut* to fill what they saw as a political void, each chipping in to pay the $1,000 cost of every issue. Frustrated by grad school, John left Indiana without finishing his degree and joined many other young people heading west to San Francisco. He got classes to teach at Sonoma State, San Francisco State and the East Bay Socialist School. He assembled a Berkeley *Jump Cut* collective. Chuck says, "While there were scheduled weekly meetings, in fact the house was a 24/7 locus of political filmmakers, critics and students, people passing through town who carried on a running dialog about politics, media issues and life in general." He went to Cuba, to Nicaragua, El Salvador and Mexico, meeting film-makers. He did an extensive set of interviews with East German film-makers. His political activism doubled as research for his teaching and writing for *Jump Cut*.

However, none of this got him a full-time job. His wife left him, taking their son Andy back to Germany. John found himself drawn into the union and increasingly active on behalf of other Lecturers. Then in the 1990s, like a lot of other people, he was laid off. Luckily, he had sensed the tightening of the job market, and had contacted his department at Indiana and finished his dissertation, so he was now able to land a tenure line job—but it was in Ithaca, New York, so he and his second wife, Gail, did the every-other-month marriage commute between California and New York. His colleague at Ithaca, Patricia Zimmerman, describes his intuitive organizer behavior there: "John would saunter down the halls of the third floor ... popping into faculty offices to chat about the news or campus gossip. He would hang out in the mail room, talking passionately about books he had just read, movies he had seen, and national and international politics that outraged him. He would always ask colleagues what they were reading." But when Gail got a full-time job in Washington, DC working for the new reform president of the Teamsters, Ron Carey, John quit Ithaca and moved to DC. This lasted only a few months because Carey was ousted. They still had their house in Oakland, so they moved back.

In spite of the ongoing cutbacks, John got a class at San Francisco State. By this time the union, the CFA, had carried out significant changes. John first reluctantly and then actively got involved again. Within a year, the CFA hired him as staff, something worth noting because many unions prefer to hire staff with other union staff experience, not people from the workforce. This was a job with full benefits and a pension, something nearly all contingents will appreciate. His personal history as a Lecturer gave him the perspective from which to recognize importance of "the end of the tenured gaze." He was able to retire after about ten years as a staffer. Within a few years, however, he was diagnosed with Parkinson's. This was around the same time that he and Joe Berry were meeting to talk about organizing and said, "Let's write a book about this."

The point of summarizing John's life this way is to contrast his steady, continuous commitment to left politics in both academic and union work with his unpredictable zig-zag job history. People would say "he had a very strong moral compass." It also reveals how little he thought of himself (or was seen) as a leader or even had much name recognition. Yet every time a Lecturer working in the CSUs cashes a paycheck or plans the next year's classes, they owe thanks to John and the many other invisible activists who made it possible.

Notes

INTRODUCTION

1. This happened in January 2021, at the six state universities in Kansas. See www.insidehighered.com/news/2021/01/22/firing-professors-kansas-just-got-lot-easier.

2. A 2020 Eurofund report found significant increases in the following "new forms of employment": ICT-based mobile work, platform work, casual work, job-sharing, employee-sharing, portfolio work, voucher-based work, and interim management (leasing out). All of these can be applied to academic work. See Eurofund (2020). *New Forms of Employment: 2020 Update*. New Forms of Employment Series. Luxembourg: Publications Office of the European Union.

CHAPTER 1   STUDENT STRIKES AND UNION BATTLES

1. The 2006 film *Walkout*, directed by Edward James Olmos, depicts another student strike that same year (1968), this one involving students in five high schools in East Los Angeles. The San Francisco State strike also sparked student strikes at San Jose State and UC Berkeley.

2. Boyle, K. (1970). *The Long Walk at San Francisco State and Other Essays* (New York: Grove Press).

3. In other states and at other times, down to the present, the question of whether contingent faculty would be in the same bargaining unit as tenure-line faculty has been decided in different ways.

4. Only wage workers could be Wobbly (Industrial Workers of the World, or IWW) members, but this included any and all wage workers, including prostitutes. However, bankers, lawyers and gamblers could not be members, an exclusion carried over from the Knights of Labor.

5. The historic debate of craft versus industrial union organizing—and from which the IWW took the term "industrial"—meant organizing all workers of a particular employer regardless of what they did (what craft they practiced) and by extension, all workers in a particular industry—such as mining, steel manufacturing, healthcare, or education. Part of the resistance to industrial organizing came from the fact that it implied organizing Black and white workers into the same union.

6. In the healthcare field, a comparable discussion is referred to as "scope of practice."

7. The first union of teachers was founded in 1897 in Chicago by Margaret Healy, an elementary school teacher. They joined the Chicago Federation of Labor in 1902 and became Local 1 of the AFT in 1916 when the AFT was formed.

8. "Act like a union" is a familiar phrase used in organizing. It means that the group carries out actions that a union would carry out, even without legal recognition, such as representing members in grievances, speaking out in public forms, engaging in collective action up to and including strikes, negotiating to the extent possible and engaging in legislative action.

9. There are today 23 CSU campuses with about 27,000 faculty. A "statewide unit" meant that faculty on all campuses would be in one unit.

10. Anderson, G., and Stulic-Dunn, H. (undated, approximately 1978–79). Origins of UPC's Committee on Lecturers, from *UPC Lecturers' Newsletter #2*.

11. Al Shanker had famously led the strike against community control of the schools and teacher hiring in New York in 1968. "Community control" would have meant the hiring of more Black educators to reflect the composition of the community and the students. Shanker was publicly opposed to affirmative action on principle.

12. Decertification is the legal process by which a certified union has its legal recognition withdrawn as a result of a vote by the bargaining unit.

13. Kurzweil's recollection is likely to be accurate, but most people just say "less than 100."

14. Norm Swensen, longtime president of the AFT 1600 local in the City Colleges of Chicago and a national leader in the AFT higher education council, told the authors that the reason the UPC lost the election was because they had "crazy chapter presidents like Kurzweil, the Communist."

15. Kurzweil expected that this suggestion coming from himself would have been met with resistance because, while he was a strong and effective local union leader, he was also known to be a member of the Communist Party which would add a degree of shock value to anything he proposed.

CHAPTER 2   LAYOFFS AND HARD YEARS FOR ORGANIZING

1. Interview with Elizabeth Hoffman, March 31, 2014, Long Beach, California.

2. Even in 1980, tenured faculty members at private institutions really did not have sufficient managerial authority in the academic and hiring areas to justify a decision of legal managerial status. Since then, they have lost much of the even advisory power that they had in 1980, making the *Yeshiva* decision even more contrary to present reality.

3. PATCO struck the air traffic control system in 1981, after Reagan reneged on his promise to them to staff the system more adequately in exchange for their endorsement in the 1980 election. Reagan responded to the strike by firing all PATCO members (who were federal employees) and brought in military air traffic controllers. The importance of the strike was that first, Reagan sent a message to employers that strike-breaking of the most extreme sort was now

on the table, and second, it demonstrated that the rest of the labor movement would not militantly mobilize at a level necessary to defend all workers.

4. A "policy" is not binding; it can be implemented or ignored by the party that has power in any given situation. Thus some people were "out" and others weren't.

5. Interview with Nina Fendel, November 18, 2019, Berkeley, California.

6. Prop 13, the beginning of the "tax revolt" of the new right-wing ascendancy, restricted property tax increases to 1 percent per year, on both residential and commercial properties. It also required a two-thirds majority vote to pass any tax bill at the state or local level, as well as a two-thirds majority vote to pass the annual state budget, which gave the Republicans a budget veto that would last thirty years.

7. The purpose of this would be to display publicly the inadequate working conditions of Lecturers.

8. Interview with Jane Kerlinger, December 9, 2019, Geyserville, California.

9. Mediation is a process by which an outside "mediator" attempts to negotiate an agreement while moving back and forth between two parties in a dispute, the goal being to gain voluntary agreement. Arbitration is generally a process by which both parties agree to submit a dispute to the decision of an arbitrator, who renders an "award" (decision). There are many styles of arbitration proceedings, even including some few that are not binding on the parties but only advisory. In binding grievance arbitration that interprets a particular section of a collective bargaining agreement, the arbitrator's award functions like judicial review of legislation and is therefore binding on future interpretations of that particular contract article.

10. Teamsters for a Democratic Union (TDU) was the most successful of the many reform movements begun by radicals in the 1970s in various unions. These reform movements were partly in response to the same political forces, both inside the unions and from the employers, that represented the neo-liberal offensive and opened the door to casualization in higher education.

11. The Teamsters were still under government supervision following a consent decree from a corruption case.

12. A full explanation of this case can be found in Crowe, K. The Vindication of Ron Carey, *Union Democracy Review 2001–2002* (December–January), www.uniondemocracy.org/UDR/22-vindication%20of%20Ron%20Carey.htm (accessed April 29, 2021).

13. CNA is the largest union of registered nurses. It would later go on to organize across the country as National Nurses United (NNU/AFL-CIO) and make a significant impact on electoral politics, especially fighting for Medicare For All and supporting the Bernie Sanders presidential campaigns.

CHAPTER 3 REVOLUTION IN THE UNION

1. Interview with Nina Fendel, November 18, 2019, Berkeley, California.

2. Interview with Jane Kerlinger, December 9, 2019, Geyserville, California.

3. This is information for readers who may be contemplating doing this. Current electronic versions of remote signings create both dangers and opportunities.

4. Interview with Susan Meisenhelder, March 21, 2014, Huntington Beach, California. As Elizabeth Hoffman explained, individual salary increases were supposed to be "step" increases—automatic depending on the number of years you had worked. "Merit" increases were discretionary for the administration to award and therefore likely to reflect favoritism rather than "merit." The fight against "merit pay" would never be over.

5. There were also polarizing debates about affirmative action going on at that time. The US Supreme Court *Bakke* decision, which limited the application of affirmative action in California in higher education, was 1978. It was important for a progressive union to have some leadership of people of color.

6. This was a situation where the union's threat of reverting back to civil service regulations could be raised.

7. Terry Jones had just completed the first term of what was traditionally a two-term presidency. It was not expected that someone would run against him after his first term. Moving a contested election would mean that a Black president, if he lost, would be atypically only a one-term president.

8. Tim Sampson was Professor of Social Work at SF State and a long-time activist, known and trusted by many people.

9. It was only one local, but was a statewide local representing at least 23,000 faculty on 23 campuses.

10. Interview with Bob Muscat, August 25, 2016, San Francisco, California.

11. The triangle is the Greek letter delta, also the symbol for change, so the symbols mean "change over time equals leverage."

CHAPTER 4 "THEY HAVE NOTHING TO TEACH US"

1. Unions have responded to *Janus* with intensified organizing, resulting in actual increases in numbers of members and union finances: Rainey, R., and Kullgren, I. (2019), 1 year after Janus, unions are flush, *Politico* (May 17), www.politico.com/story/2019/05/17/janus-unions-employment-1447266 (accessed May 7, 2021).

2. Saul Alinsky is recognized as one of the founders of modern community organizing, starting in the 1930s in the South Side of Chicago. His theories are still controversial. He wrote the most famous biography of John L. Lewis, founder of the CIO. Bob Muscat was also influenced by Alinsky.

3. A quote from convicted felon Ivan Boesky, fictionalized as Gordon Gekko in the movie *Wall Street* with Michael Douglas. In 1986, Boesky gave a speech at Berkeley in which he said, "Greed is good."

4. Jonathan Karpf (interviewed on April 15, 2020) remembers Lecturers' Council meetings with only five or six Lecturers present.

5. Burning Man is an annual retreat in the desert above Reno, Nevada, that combines a pop-up city with extreme art and attracts thousands.
6. This is a labor education tool known as a SWOT analysis, because of the first letters of each word in the name. It is used as a strategic planning device.
7. Jeff Lustig died in 2011.
8. For well-known Ruckus Society actions since 1999, when they appeared on the world stage through their trainings and actions at the Seattle World Trade Organization demonstration, see https://ruckus.org/about-us/mission-history/ (accessed January 14, 2021).

CHAPTER 5 FOUR TRANSITIONS AND HOW CASUALIZATION SERVED MANAGERS

1. Rhoades, G. (2009). Carnegie, DuPont Circle and the AAUP: (Re)Shaping a Cosmopolitan Locally-Engaged Professoriate, *Change Magazine*, January–February (author print, no page available).
2. This is a controversial argument. See Karabel, J. and Brint, S. (1989). *The Diverted Dream: Community Colleges and the Promise of Educational Opportunity in America, 1900–1985* (New York: Oxford University Press).
3. Today, this remarkable fact is met with incredulity by people applying for jobs in higher education. But in 1977, this was recognized in the *Report of the Task Force on Temporary Faculty to the Office of the Chancellor of the State Universities and Colleges* (Office of the Chancellor, California State University System): "During the sixties, anyone coming out of a doctoral program could routinely expect to obtain a probationary appointment. This is no longer the case" (p. 9). The authors do not footnote this as it was common knowledge at the time.
4. This document is usually referred to as "The AAUP Statement," but it is actually the joint *Statement of Principles on Academic Freedom and Tenure, by AAUP and the Association of American Colleges*, thus making it a labor-management agreement: www.aaup.org/file/1940%20Statement.pdf (accessed December 30, 2020).
5. As part of the re-emergence of Marxism and interest in class analysis in the 1960s and 1970s, Erik Olin Wright pioneered the concept of contradictory class location as a way of describing the location, behavior and consciousness of the increasing number of people who did not fit easily into traditional class analysis. See Wright, E.O. (1976). Class Boundaries in Advanced Capitalist Societies, *New Left Review*, No. 98 (July–August): 3–41.
6. One part of Freedom Summer was the Freedom Schools, a curriculum that combined literacy, preparation for citizenship and general education, to fill in the blanks left by underfunded segregated schools. In later years, the experience of Freedom Summer teachers and that particular curriculum became influential in graduate schools of education. The names most often associated with this project are Bob Moses and Staughton Lynd.

7. Cane, R.A. (April 28, 1977) (memo). TO: The Chancellors Office and the Part Time Temporary Task Force of the Chancellor's Office. UPC Lecturers' File, San Francisco State Labor Archives. This memo, addressed to the Chancellor's Office/Part Time Temporary Task Force, contributed to the December 1977 *Report of the Task Force on Temporary Faculty.*

8. The Powell Memo was a 1970 memo to the US Chamber of Commerce laying out a strategy for how business interests could be reasserted politically and economically. It is known by the name of its author, future Supreme Court Justice Lewis Powell. It is often seen as one of the founding documents of neo-liberalism in the United States.

9. *Report of the Task Force on Temporary Faculty of the Office of the Chancellor,* California State University and College, December 1977 (p. 6).

10. Referring to the overall growth of temporary appointments and whether this stemmed from a general plan or not, the December 1977 *Report of the Task Force on Temporary Faculty* states: "This has not happened as a direct response to any central directive or overall plan. It results from decisions taken more or less independently on the 19 campuses" (p. 7).

CHAPTER 6 BLUE SKY #1 ORGANIZING AND ECONOMICS

1. Two aspects of "acting like a union" are pursing grievances through whatever channels are available, and pushing for legislative changes. An example from the unrecognized non-collective bargaining days in the CSU system is that United Professors of California (UPC), as early as 1975, pursued successfully a legislative change to mandate access to the existing legal civil service grievance procedure (which was in fact pretty good, ending in binding arbitration). This access had been denied to Lecturers as temporary employees. This change came about because of the union's attempt to pursue grievances for all its members. All of this was before recognized collective bargaining law or recognition of the union. See *UPC Part Time Temporary Faculty Newsletter,* Fall 1975, no date (possibly first issue).

2. The classic book on this is Walton, R.E., and McKersie, R.B. (1965). *A Behavioral Theory of Labor Negotiations* (New York: McGraw-Hill).

3. These laws include the Windfall Elimination Provision and the Government Pension Offset.

4. Berry, J., Stewart, B., and Worthen, H. (2007). *Access to Unemployment Insurance Benefits for Contingent Faculty* (Chicago, IL: COCAL).

5. Worthen, H., and Berry, J. (1999). *Contingent Faculty in Public Higher Education in Pennsylvania, Spring 1999: Focus on Community Colleges.* A report presented at the annual meeting of the Faculty Coalition on Higher Education, April 30–May 1, 1999, Keystone Research Center, 412 North Third Street, Harrisburg, PA 19101.

CHAPTER 7 BLUE SKY #2 JOB SECURITY, ACADEMIC FREEDOM AND THE COMMON GOOD

1. Most conflicts like this involving non-tenure line faculty are not publicized and may even include a non-disclosure agreement as part of the settlement. Giles' case was publicized because there was a union at Roosevelt, which took up the case. Joe Berry was Giles' shop steward at Roosevelt when it occurred.
2. See *Bargaining for the Common Good* at: www.bargainingforthecommongood. org (accessed April 29, 2021).
3. More Effective Schools (MES) was an explicitly anti-racist AFT-branded campaign that called for schools to be better integrated into their surrounding communities and equalized in resources, with maximum teacher-staff-and parent input. It was very similar to the Community Schools initiative being pursued in the teacher union movement today but with a strong anti-racist component. It was discontinued after Albert Shanker became AFT President in 1974.
4. Brady, M. (2019). Workers and Renters of the World Unite! *Jacobin*, September 13, https://jacobinmag.com/2019/09/rent-control-affordable-housing-teacher-pension-funds-evictions (accessed January 15, 2021).

CHAPTER 8 BEYOND THE SAUSAGE-MAKING: A CLOSE LOOK AT THE CFA-CSU CONTRACT

1. The 2018 study by LeFlore confirms this in a comparison between the CFA-CSU contract with the PSC-CUNY contract: LeFlore, M.A. (2018). Contingent Faculty Unionization: The Impact of Collective Bargaining on Course Pay, Benefits, and Contract Lengths for Part-time Faculty at Four-year Institutions in the US. MA Capstone Paper in Policy Studies, Washington ResearchWorks Archive. http://hdl.handle.net/1773/42753 (accessed April 22, 2021).
2. The CFA contract is available on the web at www.calfac.org/resource/collective-bargaining-agreement (accessed April 29, 2021). The CFA faculty handbook is available at www.calfac.org/Lecturers-council (accessed April 29, 2021).
3. This is an example of a formal rule that is almost never enforced. Rules like this can functions as "gotchas," but they can also be the subject of past practice grievances if the employer has failed to enforce them in the past.
4. We note this choice specifically because of our experience in the labor relations of workers in Vietnam where, at least in 2017, union representatives were by law paid directly by the employer. That contributed to loyalty to the employer on the part of the representative and lack of trust in the representative by the workers.
5. In 2020, there were 29,000 faculty, of whom 17,000 were Lecturers, of whom 83 percent were part time.
6. The term "Range" is used in the CSU system to mean what is ordinarily referred to as "column" in most faculty contracts. In most contracts, you are placed in a

column by your qualifications and then you progress in steps based on seniority or sometimes "merit."

7. Job position titles and pay in the CSU system, modified for this book from the tables provided on pp. 48 and 49 of *The CFA Lecturer's Handbook Through 2020*, at https://www.calfac.org/sites/main/files/file-attachments/2018_cfa_lecturers_ handbook_web_0.pdf (accessed May 2, 2021). Thanks to Jonathan Karpf for guiding us through this topic.

8. See *Cervisi vs Unemployment Insurance Appeals Board* (February 1, 1989). Court of Appeal, First District, Division 4, California: https://caselaw.findlaw. com/ca-court-of-appeal/1758049.html (accessed April 29, 2021).

9. We write this part in the middle of the Covid-19 virus pandemic; the budget meltdown may have arrived.

CHAPTER 9 STRATEGIES EMERGING FROM PRACTICE

1. See Berry, J. (2005). *Reclaiming the Ivory Tower, Organizing Adjuncts to Change Higher Education* (New York: Monthly Review Press), and our chapter, The Metro Strategy: A Workforce-Appropriate, Geographically Based Approach to Organizing Contingent Faculty, in I. Minune (2018) (ed.), *Contextualizing and Organizing Contingent Faculty* (Lanham, MD: Lexington Books).

2. Their story is told in Vinnie Tirelli's dissertation (2008), The Invisible Faculty Fight Back: Contingent Academic Labor and the Political Economy of the Corporate University; Unpublished dissertation submitted to the Graduate Faculty in Political Science in partial fulfillment of the requirements for the degree of Doctor of Philosophy, The City University of New York, NY.

3. This is a reference to INCITE! Women of Color Against Violence (2017). *The Revolution Will Not Be Funded: Beyond the Non-Profit Industrial Complex* (Durham, NC: Duke University Press).

4. In 2006, a series of walkouts took place in cities in the US under the title "A Day Without Immigrants"; they were intended to show how essential the work of immigrants is to the economy. The walkouts culminated in massive demonstrations on May Day. A popular 2004 satirical movie, *A Day without a Mexican*, also made the same point.

5. See Rein, M., Legion, V., and Ellinger, M. (2021). *Free City! The Fight for San Francisco City College and Education* (Oakland, CA: PM Press).

CHAPTER 10 THE CONTINGENT FACULTY MOVEMENT AS A SOCIAL MOVEMENT

1. Blunden, A. (2019). *Hegel for Social Movements* (Leiden, The Netherlands: Koninklijke Brill NV); all quotes from Blunden in this chapter are from p. 61 of this volume.

2. See https://youtu.be/cXXYsVhlv1U for a view of the small spinning wheel used by Gandhi and an example of how it was (and is still being) used as a symbol of a movement (accessed January 18, 2021).

3. We contrasted two views of "quality"—one judged by typical outcome measures from a management perspective and the other from a student and faculty learning perspective—in our article, "A CHAT analysis of college teaching": Worthen, H., and Berry, J. (2006). Our Working Conditions are Our Students' Learning Conditions: A CHAT Analysis of College Teaching (pp. 145–159). In Sawchuk, P., Duarte, N., and El Hammoumi, M. (eds.), *Critical Perspectives on Activity: Explorations Across Education, Work & Everyday Life*. Cambridge, UK: Cambridge University Press.

CHAPTER 11 WHAT GETS PEOPLE MOVING?

1. See Myles Horton's *The Long Haul, an Autobiography*, as told to Kohl, J. and Kohl, H. (1990) (New York: Doubleday/Dell).

2. Parker, M., and Gruelle, M. (1999). *Democracy is Power: Rebuilding Unions from the Bottom Up* (Detroit, MI: Labor Notes).

3. This is a collective action-version of David Kolb's Learning Cycle. It turns the learning cycle into an agenda for a meeting. It is also described in Burke, B., Geronimo, J., Martin, D., Thomas, B., and Wall, C. (eds.) (2002). *Education for Changing Unions* (Toronto: Between the Lines).

4. For a general background to these ideas, refer to the classic work on the Haitian Revolution: James, C.L.R. (1938). *The Black Jacobins: Toussaint L'Ouverture and the San Domingo Revolution* (London: Secker and Warburg).

5. As Andy Blunden explains in *The Origins of Collective Decision-Making* (Chicago, IL: Haymarket, 2017), the very idea of democratic decision making arose from the needs of craftsmen and artisans to organize themselves democratically on the basis of their occupation. This is what made them dangerous and illegal, because it undermined the ideology of feudalism.

6. Note the founding slogan of the AFT, "Education for Democracy and Democracy in Education."

7. See Worthen, H. (2015). Organizing as Learning: The Metro Strategy and a Community of Practice for Faculty, *WorkingUSA: The Journal of Labor and Society*, Vol. 18, No. 3: 421–434.

CHAPTER 12 WHO IS THE ENEMY AND WHO ARE OUR ALLIES?

1. In November 2019, Matt Bevin, the Republican governor of Kentucky was defeated by a Democratic challenger Andy Beshear, but Bevin made last-minute appointments of some of his favored aides to various positions in the governance regime of the state's higher education system. This use of higher education policy-making position as a sinecure for the favored (or the disfavored

who can't be fired) has a long history in many states. (see *Chronicle of Higher Education*, November 26, 2019).

2. This can be called a union's coalition operating strategy. M.A. LeFlore identified it along with having a combined bargaining unit (tenure-line and contingent faculty) as being a significant influence on positive outcomes: "… with higher course pay, greater odds for health and retirement benefits, and longer contract terms generally associated with unions that include all types of faculty as members, rather than representing part-time instructors separately from others, and that choose to ally with other unions in their efforts, rather than operate alone": LeFlore, M.A. (2018). Contingent Faculty Unionization: The Impact of Collective Bargaining on Course Pay, Benefits, and Contract Lengths for Part-time Faculty at Four-year Institutions in the US. MA Capstone Paper in Policy Studies, Washington ResearchWorks Archive. http://hdl.handle.net/1773/42753 (accessed April 22, 2021).

3. See Blanc, E. (2019). *Red State Revolt: The Teachers' Strike Wave and Working-Class Politics* (London: Verso).

4. Caref, C. et al. (2014). *The Schools Chicago's Students Deserve 2.0* (Chicago, IL: Chicago Teachers Union AFT Local 1): www.ctulocal1.org/reports/schools-chicagos-students-deserve-2/ (accessed January 2, 2020).

5. The concept of "distortions of human development under capitalism" depends on looking at human development as occurring within a social, historical and cultural framework—not just the development of individuals on their own or within a family or even a school, but within a society. Specifically, we mean psychological and cognitive disabilities ranging from lack of empathy, envy, despair, alienation and bullying to obesity, eating disorders and stress-related auto-immune illnesses. Relating these to the social situation in which they develop is a concept drawing from Vygoskian psychology. See Sultanova, A., and Ivanova, I. (2011). The Features of a Social Situation of Development of Children under Modern Russian Conditions, *Tätigkeitstheorie: E-Journal for Activity Theoretical Research in Germany*, No. 4: 53–64.

6. See Herbert, W.A., and Apkarian, J. (2019). You've Been With the Professors: An Examination of Higher Education Work Stoppage Data, Past and Present, *Employee Rights and Employment Policy Journal*, Vol. 23. The Appendix contains a table of 43 work stoppages. Among grad assistants (7), Tenure Line (1), TT-NTT (8), Non-Academic (21) and NTT (6).

CHAPTER 13 WHAT IS "PROFESSIONALISM" FOR US?

1. See Wright, E.O. (1976). Class Boundaries in Advanced Capitalist Societies. *New Left Review*, No. 98 (July–August): 3–41, and *idem* (1978). *Class, Crisis and the State* (New York: Schocken Books).

2. Here we are acknowledging a debt to the gay rights movement, not just appropriating their expression. We mean to reflect how steep is the climb required to

achieve a major change in consciousness. The debt that we need to acknowledge is that the gay rights movement taught us that, for people whose second-class status can be hidden individually, we cannot gain public, organized allies until we make our collective existence public and organize ourselves to demand first-class status.

3. In 1941, AFT Local 2, referred to as "the Teachers' *Union*," in New York City was purged (that is, their charter was revoked by the National AFT) for being too left-wing and too close to the CIO. The AFT then chartered a new local and called it "Local 2." That new local called itself The Teachers' *Guild*, in part as a way to distance itself from the old "union."

4. See the biography by Kate Rousmaniere (2005), *Margaret Haley, Citizen Teacher: The Life and Leadership of Margaret Haley* (New York: SUNY Press), and the general history of teacher unions by Marge Murphy (1990), *Blackboard Unions: The AFT and the NEA, 1900–1980* (Ithaca, NY: Cornell University Press).

5. "Community of practice" is a reference to the work of Jean Lave, whose work with Etienne Wenger established this concept as a way to think about learning as a socially situated activity; see Lave, J., and Wenger, E. (1991). *Situated Learning: Legitimate Peripheral Participation* (Cambridge, UK: Cambridge University Press).

6. Chicago Teachers' Union (2018), *The Schools Chicago's Students Deserve*: https://news.wttw.com/sites/default/files/Chicago%20Teachers%20Union%20 report_0.pdf (accessed April 29, 2021).

7. *Survey by the AFT Task Force on Part-Time College and University Faculty of Local Union Leaders* (1979), attached with a memo to Warren Kessler, President [of the] UPC, August 1979, date January 9, 1979 from James Ward, Director of Economic Research, AFT, Washington, DC. No final formal report was ever made.

8. AAUP Council (November 2012). *The Inclusion in Governance of Faculty Members Holding Contingent Appointments*, AAUP Statement on Contingent Faculty in Shared Governance Bodies. www.aaup.org/report/inclusion-governance-faculty-members-holding-contingent-appointments and www.aaup.org/system/files/members/Governance-Inclusion-Contingent.pdf# overlay-context=report/governance-inclusion (pdf) (accessed May 4, 2021). Statistical information in the report was updated in 2014.

9. The strategic alternative is exemplified by the situation at University of San Francisco, a Jesuit institution, where they incorporated all shared governance activities into the union contract, combining Senate and union power and responsibilities in one body. The authors only know this one example of this kind of arrangement.

CHAPTER 14 HOW DOES IT FEEL?

1. It is also useful when the organizer can show how capitalism, especially in the form into which it has developed in the US in the 2000s, shapes individual

human development, emotional, cognitive and physical, into distorted, self-destructive patterns that can only be corrected in a collective context.

2. During organizing, employees are usually subjected by management to one-on-ones with supervisors or anti-union group meetings in which the worst possible stories, including many that are untrue, about unionization are presented as a way to discourage organizing. Warning workers in advance about what they will hear is called "inoculation."

3. This is described in Hoeller, K. (ed.) (2014). *Equality for Contingent Faculty: Overcoming the Two-Tier System* (Nashville, TN: Vanderbilt University Press, pp. 9–27).

4. Kathleen Christianson's chapter "Toiling for Piece Rates and Accumulating Deficits" is survey-based research with hiring officers in higher ed; she reports their stated opinion that, after three years, experience as a contingent is a deficit rather than a asset in hiring: Barker, K., and Christianson, K. (eds.) (1998). *Contingent Work: American Employment Relations in Transition* (Ithaca, NY: Cornell ILR Press).

5. https://teachingamericanhistory.org/library/document/what-to-the-slave-is-the-fourth-of-july/ (accessed April 22, 2021).

6. See Weingarten, K. (2004). *Common Shock: Witnessing Violence Every Day*, New York: New American Library.

CHAPTER 15 "IS THIS LEGAL?"

1. Montana is the only state where the default is not employment-at-will.

2. Because of the recent changes in government administration and therefore the NLRB and the Supreme Court, all of these interpretations and rules are likely to be subject to further litigation, controversy and new rule making going forward. A good summary can be found in Jaschik, S. (2015). A Big Union Win, *Inside Higher Ed* (January 2), www.insidehighered.com/news/2015/01/02/nlrb-ruling-shifts-legal-ground-faculty-unions-private-colleges (accessed May 5, 2021).

3. See http://catholiclabor.org/catholic-employer-project/.

4. The actual opinion is at www.supremecourt.gov/opinions/17pdf/16-1466_2b3j.pdf (accessed May 3, 2021); see also Semuels, A. (June 27, 2018). Is This the End of Public Sector Unions in America? *The Atlantic*, www.theatlantic.com/politics/archive/2018/06/janus-afscme-public-sector-unions/563879/ (accessed May 3, 2021).

5. Public-sector collective bargaining laws by state appear in lists on the Internet such as https://education.findlaw.com/teacher-rights/teacher-s-unions-collective-bargaining-state-and-local-laws.html.

6. Sometimes legislative action can bear fruit years later. In an example from the CSU history, even before the state collective bargaining law was passed, the United Professors of California (UPC) succeeded in sponsoring a change in

state law that defined an offer of employment as "not reasonable assurance," if contingent upon enrollment, funding, or program needs. This statute then became, years later, the basis for the statewide *Cervisi* lawsuit (*Cervisi vs Unemployment Insurance Appeals Board*, 1989) that established the principle for the right to unemployment benefits for all contingent faculty.

7. The law known as "the 67% law" was a state law limiting the employment of part-time temporary faculty in the California community colleges to 67% of a full time load. The idea was to prevent filling faculty positions with people hired to teach a full load but only receive pay at the part-time temporary rate and not have continuity of employment (job security). Originally, the 67% was 60%, but the limit was raised to allow people to teach two-thirds of a load. The consequence of hiring someone for three consecutive semesters to teach as a part time temporary but carry over 60% (and then later, 67%) was that their assignment would "trigger" the tenure process and lead potentially to being hired full-time at a tenure-line rate with job security. The contradictions displayed by this attempt to balance common sense and expediency were and still are dazzling. There were a few cases where someone actually did gain a full-time faculty position as a result of this process. One famous case took place in the Peralta district in Oakland, where several part-time faculty went through this process and were awarded with "permanent" jobs, never to be expanded beyond 60%.

8. Two recent books that make this point are Joe Burns' *Strike Back! Using the Militant Tactics of Labor's Past to Re-ignite Public Sector Unionism Today* (Brooklyn, NY: IG Publishing & Workrights Press, 2014), and Robert M. Schwartz's *No Contract, No Peace!* (Boston, MA: Workrights Press, 2012 [first published in 2006 as *Strikes, Picketing and Inside Campaigns*].

9. Schwartz, R.M. (1999, 2003). *The Labor Law Source Book: Texts of Federal Labor Laws* (Cambridge, MA: Workrights Press). Schwartz is also the author of indispensible guides to grievance handling, enforcement of specific laws like the FMLA (Family Medical Leave Act, 1993) and strikes.

10. The Labor Management Relations Act of 1947, better known as the Taft–Hartley Act, is a United States federal law that restricts the activities and power of labor unions. It was a reaction to the wave of organizing up to, during and after World War II and the strike wave after the war. Taft-Hartley required that any official of a union using the services of the National Labor Relations Board would have to submit an affidavit of non-Communist membership by the union's officials even if they were democratically elected. It also 1) allowed states to enact right-to-work laws which outlawed union shop and mandatory union dues; 2) outlawed many strike tactics such as mass picketing, secondary boycott, sympathy strikes and hot cargo agreements; 3) outlawed foremens' unions and other unions of supervisors, 4) established the category of unfair labor practices by unions and allowed employers to file ULPs against unions, and 5) allowed employers free speech rights during union organizing, which opened the door to captive

audience meetings and other kinds of employer intervention in union election campaigns using tactics that were not available to the unions. In summary, it outlawed most of the union tactics and actions that had been used to successfully organize a major section of the working class in the 1930s and 1940s. The tricks and traps of Taft-Hartley are good evidence of why "Is this legal?" is not the first question an employee group should ask when considering what to do.

CHAPTER 16 WHAT ABOUT LEFTISTS?

1. Bradbury, A., Brenner, M., Slaughter, J., Brown, J., and Winslow, S. (2014). *How to Jumpstart Your Union: Lessons from the Chicago Teachers* (Detroit, MI: Labor Notes).
2. Good historical descriptions of this are in Robin D.G. Kelley's *Hammer and Hoe: Alabama Communists during the Great Depression* (Chapel Hill, NC: University of North Carolina Press, 1990), and Toni Gilpin's *The Long Deep Grudge* (Chicago, IL: Haymarket Books, 2020).
3. As is well known, this is what happened in the US labor movement to union leaders and staff who were known communists after World War II.
4. Salting is a labor union tactic involving the act of getting a job at a specific workplace with the intent of organizing a union. A person so employed is called a "salt."
5. Hanson, C., Paavo, A., and Sisters in Labour Education (2019). *Cracking Labour's Glass Ceiling: Transforming Lives through Women's Union Education* (Black Point, Nova Scotia: Fernwood Publishing).
6. See Worthen, H. (2014). *The Status of Labor Education Programs in Higher Education in the US* for a report on the various kinds of attacks on labor education programs: https://uale.org/resource/links/ (accessed April 22, 2021).

CHAPTER 17 HOW DO WE DEAL WITH UNION POLITICS?

1. Examples would be bargaining over tenure-line positions (either the number or the percentage), hiring, upgrading, how the employer will respond to questions from unemployment benefits offices, how the employer would use money that might come from future legislation, whether or not an employer would take a position on a legislative issue, etc. In the case of the Los Angeles UTLA teachers strike, they bargained to force the pro-charter school board to support state legislation restricting charter schools.

CHAPTER 18 HOPES AND DANGERS

1. "*La crisi consiste appunto nel fatto che il vecchio muore e il nuovo non può nascere: in questo interregno si verificano i fenomeni morbosi piú svariati.*" *Passato e presente, Quaderni del carcere,* "Ondata di materialismo" e "crisi di autorità," volume I, quaderno 3, p. 311, written circa 1930. English translation *Selections*

from the Prison Notebooks, "Wave of Materialism" and "Crisis of Authority" (New York: International Publishers, 1971), pp. 275–276.

2. See https://junctrebellion.wordpress.com (accessed April, 21, 2021).

3. In the US, many K–12 school districts have allowed the creation of schools that are set up separately ("chartered") from the main districts. They may be set up in a variety of ways, for example, by individuals who have a particular idea of how to run a school, by collectives, by community groups, or by companies that own chains of schools set up this way. They can be for-profit chains or non-profits. In many cases, the charter schools are exempt from the union contracts that cover teachers, are free to select their students, can develop their own curricula, and otherwise differentiate themselves from the main body of the school district. They nonetheless are supported by public funding. One origin of charter schools in the US was in the whites-only "Christian academies" set up in the South to escape de-segregation after *Brown vs Board of Education*, the 1954 Supreme Court decision that said that separate was not equal and so separate Black and white schools violated the Constitution. See MacLean, N. (2017), *Democracy in Chains: The Deep History of the Radical Right's Stealth Plan for America* (New York: Viking) for details. Ironically, the other origin of charters was AFT President Albert Shanker, who coined the term and had the idea that they could be experimental schools, started and run by teachers, as pilot projects to advance non-bureaucratic and educational reforms. For many reasons, the political right in the 1980s saw this as an opening for massive privatization.

Bibliography

AAUP (1940). *1940 Statement of Principles of Academic Freedom and Tenure*. American Association of University Professors. www.aaup.org/file/1940%20 Statement.pdf (accessed December 30, 2020).

AAUP (2006). *Contingent Faculty Index*. Washington, DC: American Association of University Professors. Updated by Steven Shulman under the title *Faculty and Graduate Student Employment Report 2013* published by Center for the Study of Academic Labor, Colorado State University, 2017.

AAUP (November 2012). *The Inclusion in Governance of Faculty Members Holding Contingent Appointments*, AAUP Statement on Contingent Faculty in Shared Governance Bodies. www.aaup.org/report/inclusion-governance-faculty-members-holding-contingent-appointments and www.aaup.org/system/files/members/Governance-Inclusion-Contingent.pdf#overlay-context=report/governance-inclusion (pdf) (accessed May 4, 2021).

AAUP (2017). *The Adjunct Coloring Book*. Washington, DC: American Association of University Professors.

AAUP and Front Range Community College Chapter (FRCC/AAUP) (2014). The Adjunct Cookbook. Washington, DC: American Association of University Professors.

Abel, E.K. (July 1977). The Professional Proletariat: Teachers in California Community Colleges, *Radical Teacher*, July 1977. See also special section on contingent faculty in this issue.

Abel, E.K. (1984). *Terminal Degrees: The Job Crisis in Higher Education*. New York: Praeger Publishers.

Anderson, G., and Stulic-Dunn, H. (undated, approximately 1978–79). Origins of UPC's Committee on Lecturers, *Lecturer Newsletter #2*.

Anderson, J. (2019). How the Two-Tiered System in Higher Education Gets Reproduced (and Hopefully Abolished), *Counterpunch* (July 5). www.counterpunch.org/2019/07/05/how-the-two-tiered-system-in-higher-education-gets-reproduced-and-hopefully-abolished/ (accessed April 21, 2021).

Angulo, A.J. (2018). From Golden Era to Gig Economy: Changing Contexts for Academic Labor in America. In Tolley, K. (ed.), *Professors in the Gig Economy: Unionizing Adjunct Faculty in America* (pp. 3–26). Baltimore, MD: Johns Hopkins University Press.

Arnold, G.B. (2000). *The Politics of Faculty Unionization: The Experience of Three New England Universities*. Westport, CT: Bergin & Garvey.

Aronowitz, S. (2000). *The Knowledge Factory: Dismantling the Corporate University and Creating True Higher Learning*. Boston, MA: Beacon Press.

Ashby, S.K., and Bruno, R. (2016). *A Fight for the Soul of Public Education: The Story of the Chicago Teachers Strike*. Ithaca, NY: Cornell ILR Press.

Autonomedia (2009) *Toward a Global Autonomous University: The Edu-Factory Collective*. www.autonomedia.org. Brooklyn, NY: Autonomedia.

Baldwin, R.G., and Chronister, J.L. (2001). *Teaching Without Tenure: Policies and Practices for a New Era*. Baltimore, MD: Johns Hopkins University Press.

Barker, K. and Christianson, K. (eds.) (1998). *Contingent Work: American Employment Relations in Transition*. Especially: Christianson, K., Toiling for Piece Rates and Accumulating Deficits: Contingent Work in Higher Education (pp. 195–220). Ithaca, NY: Cornell ILR Press.

Barrone, J.T. (2020). The Proletariat in Higher Education: An Introduction of Contingent Faculty as the Precarious Class (Doctoral dissertation, Northern Arizona University).

Benjamin, E. (ed.) (2003). *Exploring the Role of Contingent Instructional Staff in Undergraduate Learning*, No. 123, Fall 2003. San Francisco, CA: Jossey-Bass, Wiley.

Benjamin, E., and Mauer, M. (eds.). (2006). *Academic Collective Bargaining*. New York: The Modern Language Association, and Washington, DC: American Association of University Professors

Benson, H.W. (1979). *Democratic Rights for Union Members: A Guide to Internal Union Democracy*. New York: Association for Union Democracy.

Berry, J. (2002). Campus Equity Week: Contingent Faculty Make News, *Workplace: A Journal for Academic Labor*, Vol. 8.

Berry, J. (2005). *Reclaiming the Ivory Tower: Organizing Adjuncts to Change Higher Education*. New York: Monthly Review Press.

Berry, J. (2006). The Value of Forging Alliances, *Chronicle of Higher Education*, Vol. 52, No. 41 (June 16): B10–B11.

Berry, J. (2003). Campus Equity Week's Offspring Takes a Few Steps: Contingent Faculty Organizing in Metro Chicago, *WorkingUSA*, Vol. 6, No. 4: 32–37.

Berry, J. (2008–09). Contingent Faculty and Academic Freedom: A Contradiction in Terms. *Works and Days*, Vols. 26 and 27, No. 51/52: 359–368.

Berry, J., and Blanchette, J. (2015). Teaching and Organizing in For-Profit Higher Education: A Kaplan story, *WorkingUSA*, Vol. 18, No. 3 (September): 447–456.

Berry, J., and Worthen, H. (2012). Organizing Faculty in the Higher Education Industry: Tackling the For-profit Business Model, *WorkingUSA*, Vol. 15, No. 3 (September).

Berry, J., and Worthen, H. (2018). The Metro Strategy: A Workforce-Appropriate, Geographically Based Approach to Organizing Contingent Faculty. In Minune, I. (ed.), *Contextualizing and Organizing Contingent Faculty: Reclaiming Academic Labor in Universities*. Lanham, MD: Lexington Books.

Berry, J., and Worthen, H. (2012). Higher Education as a Workplace, *Dollars and Sense*, No. 3 (November–December): 19–23.

Berry, J., Stewart, B., and Worthen, H. (2007) *Access to Unemployment Insurance Benefits for Contingent Faculty.* Chicago, IL: COCAL.

Bischoff, C. (2019). The Spiritual Crisis of Contingent Faculty. *Journal of Moral Theology*, Vol. 8 (Special Issue 1): 115–125.

Blanc, E. (2019) *Red State Revolt: The Teachers' Strike Wave and Working-Class Politics.* London: Verso.

Blunden, A. (2017). *The Origins of Collective Decision Making.* Chicago, IL: Haymarket Books.

Blunden, A. (2019). *Hegel for Social Movements.* Leiden, The Netherlands: Koninklijke Brill NV.

Bosquet, M. (2008). *How the University Works: Higher Education and the Low-Wage Nation.* New York: New York University Press.

Bowles, S., and Gintis, H. (1976). *Schooling in Capitalist America.* New York: Basic Books.

Boyle, K. (1970). *The Long Walk at San Francisco State and Other Essays.* New York: Grove Press.

Bradbury, A., Brenner, M., Slaughter, J., Brown, J., and Winslow, S. (2014). *How to Jumpstart Your Union: Lessons from the Chicago Teachers.* Detroit, MI: Labor Notes.

Brady, M. (2019). Workers and Renters of the World Unite! *Jacobin*, September 13. https://Jacobinmag.com/2019/09/rent-control-affordable-housing-teacher-pension-funds-evictions (accessed January 15, 2021).

Burke, B., Geronimo, J., Martin, D., Thomas, B., and Wall, C. (eds.) (2002). *Education for Changing Unions.* Toronto: Between the Lines.

Burns, J. (2014). *Strike Back! Using the Militant Tactics of Labor's Past to Re-ignite Public Sector Unionism Today.* Brooklyn, NY: IG Publishing & Workrights Press.

Cadambi Daniel, M. (2016). Contingent Faculty of the World Unite! Organizing to Resist the Corporatization of Higher Education, *New Labor Forum*, Vol. 25, No. 1 (January): 44–51. Los Angeles, CA: SAGE Publications.

Cain, T.R. (2017). Campus Unions: Organized Faculty and Graduate Students in US Higher Education, *ASHE Higher Education Report*, Vol. 43, No. 3: 7–163.

Caplow, T., and Reece, J.M. (1958/1965). *The Academic Marketplace: An Anatomy of the Academic Profession … Its Mores, Its Morale, Its Machinations.* Garden City, NY: Anchor/ Doubleday Books.

Caref, C., Reese, R. , Pavlyn, J., Rothschild, S., Hilgendorf, K., Osborne, L., Johnson, J., and Geovanis, C. (2014) *The Schools Chicago's Students Deserve 2.0.* Chicago, IL: Chicago Teachers Union AFT Local 1.

Carnegie, A. (2013) *The Gospel of Wealth and Other Timely Essays.* Cambridge, MA: Harvard University Press.

Carnegie Foundation for the Advancement of Teachers. *Foundation History, the early years.* www.carnegiefoundation.org/about-us/foundation-history/ (accessed December 30, 2020).

Cervisi vs Unemployment Insurance Appeals Board (1989). Court of Appeal, First District, Division 4, California (February 1). https://caselaw.findlaw.com/ca-court-of-appeal/1758049.html (accessed April 29, 2021).

CFA (2019–20). *Lecturers' Handbook*. www.calfac.org (accessed January 28, 2021). The earliest predecessor of this was produced as the *UPC Part-Time/Temporary Faculty Handbook for CSUC Faculty, 1979–1980* by the United Professors of California Part-Time Temporary Faculty Committee.

CFA Lecturers' Council (circa 2018). *Best Practices for Lecturers, from a meeting of Lecturer members in History Dept. at SF State University*, discussion paper adopted by the CFA Lecturers' Council and SFSU History Department. No author.

Charney, M., Hagopian, J., and Peterson, B. (eds.) (2021). *Teacher Unions and Social Justice: Organizing for the Schools and Communities Our Students Deserve*. Milwaukee, WI: Rethinking Schools.

Chicago Teachers' Union (2018). *The Schools Chicago's Students Deserve*. https://news.wttw.com/sites/default/files/Chicago%20Teachers%20Union%20report_0.pdf (accessed April 29, 2021).

Childress, H. (2019). *The Adjunct Underclass: How America's Colleges Betrayed Their Faculty, Their Students and Their Mission*. Chicago, IL: University of Chicago Press.

Chomsky, N. et al. (eds.) (1997). *The Cold War and the University: Toward an Intellectual History of the Postwar Years*. New York: The New Press.

Clawson, D. (2003). *The Next Upsurge: Labor and the New Social Movements*. Ithaca, NY: ILR Press.

Coalition on the Academic Workforce (CAW) (2012). *A Portrait of Part-Time Faculty Members: A Summary of Findings on Part-Time Faculty Respondents to the Coalition on the Academic Workforce Survey on Contingent Faculty Members and Instructors*. www.academicworkforce.org/CAW_portrait_2012.pdf (accessed February 26, 2021).

Colby, R.S. (2017) *Contingency, Exploitation, and Solidarity: Labor and Action in English*. https://wac.colostate.edu/books/perspectives/contingency/ (accessed April 22, 2021).

Coles, G. (2018). *Mis-educating the Global Economy: How Corporate Power Damages Education and Subverts Students' Futures*. New York: Monthly Review Press.

Cottom, T.M. (2017) *Lower Ed: The Troubling Rise of For-Profit Colleges in the New Economy*. New York: The Free Press.

Crowe, K. (2002–02). The Vindication of Ron Carey, *Union Democracy Review 2001–2002* (December–January), https://www.uniondemocracy.org/UDR/22-vindication%20of%20Ron%20Carey.htm (accessed April 29, 2021).

Crozier, M., Huntington, S.P., and Watanuki, J. (1975). *The Crisis of Democracy: A Report on the Governability of Democracies to the Trilateral Commission*. New York: New York University Press.

Dannin, E. (2003). Organizing Contingent Academics; The Legal and Practical Barriers, *Journal of Labor and Society*, Vol. 6, No. 4: 5.

Davis, D. (2017). *Contingent Academic Labor: Evaluating Conditions to Improve Student Outcomes*. Sterling, VA: Stylus Publishing, New Faculty Majority Series.

DeCew, J.W. (2003). Unionization Among Part-time Faculty. In DeCew, J.W. (ed.), *Unionization in the Academy, Visions and Realities* (pp. 75–88). Lanham, MD: Rowan and Littlefield Publishers.

Dennis, M.J. (2021). Predictions for Higher Education Worldwide for 2021, *University World News: The Global Window on Higher Education* (January 9). www.universityworldnews.com/ (accessed February 3, 2021).

Derrickson, T. (ed.) (2003). Information University: Rise of the Education Management Organization, *Works and Days*, Vol. 21, No. 41/42.

Dewey, J. (1916) *Democracy and Education*. New York: Macmillan.

Dobbie, D., and Robinson, I. (2008). Reorganizing Higher Education in the United States and Canada: The Erosion of Tenure and the Unionization of Contingent Faculty, *Labor Studies Journal*, Vol. 33, No. 2: 117–140.

Doe, S. et al. (2011). Discourse of the Firetenders: Considering Contingent Faculty Through the Lens of Activity Theory, *College English*, Vol. 73, No. 4: 428–449.

Dougherty, K.A., Rhoades, G., and Smith, M.F. (2014). Bargaining for Part-time Contingent Faculty, *NEA 2014 Almanac of Higher Education*. Washington, DC: National Education Association.

Dougherty, K.A., Rhoades, G., and Smith, M.F. (2016). Negotiating Improved Working Conditions for Contingent Faculty, *NEA 2016 Almanac of Higher Education*. Washington, DC: National Education Association.

Douglass, F. (July 5, 1852). *What to the Slave is the Fourth of July?* Rochester, NY. https://teachingamericanhistory.org/library/document/what-to-the-slave-is-the-fourth-of-july/ (accessed April 22, 2021).

DuBois, W.E.B. (1935, 2014). *Black Reconstruction in America: An Essay Toward a History of the Part Which Black Folk Played in the Attempt to Reconstruct Democracy in America, 1860–1880*. New York: Oxford University Press.

Dubson, M. (ed.) (2001). *Ghosts in the Classroom: Stories of College Adjunct Faculty— and the Price We All Pay*. Boston, MA: Camel's Back Books.

Eagan, J. (2020). The California Faculty Association: Keeping Racial and Economic Justice at the Forefront, *Journal of Collective Bargaining in the Academy*, Vol. 11, No. 7. https://thekeep.eiu.edu/jcba/vol11/iss/1/7 (accessed April 22, 2021).

Ehrenberg, R.G. (ed.) (2004). *Governing Academia: Who is in Charge at the Modern University?* Ithaca, NY: Cornell University Press.

Elbaum, M. (2018). *Revolution in the Air: Sixties Radicals Turn to Lenin, Mao and Che*. London: Verso.

Elliott-Negri, L. (2019). Comparing Contingent Faculty Pay and Working Conditions: The City University of New York and the California State University, *PS: Political Science & Politics*, Vol. 52, No. 3: 515–516.

Eurofund (2020). *New Forms of Employment: 2020 Update*. New Forms of Employment Series. Luxembourg: Publications Office of the European Union.

European University Association (February 2021). *Universities Without Walls: Vision for 2030.* www.eua.eu/resources/publications/957:universities-without-walls-%E2%80%93-eua%E2%80%99s-vision-for-europe%E2%80%99s-universities-in-2030.html (accessed April 22, 2021).

Fabricante, M., and Brier, S. (2016). *Austerity Blues: Fighting for the Soul of Public Higher Education.* Baltimore, MD: Johns Hopkins University Press. See especially pp. 236–241.

Filipan, R. (2014). *Shouting from the Basement and Re-Conceptualizing Power: A Feminist Oral History of Contingent Women Faculty Activists in U.S. Higher Education* (Dissertation). https://www.semanticscholar.org/paper/Shouting-from-the-basement-and-re-conceptualizing-A-Filipan/2ec31fd6e1f2c3cbbb3d60130b4363ba1e827fbc (accessed April 22, 2021).

Flaherty, C. (2021). Suspending the Rules for Faculty Layoffs. www.insidehighered.com/news/2021/01/22/firing-professors-kansas-just-got-lot-easier (accessed February 3, 2021).

Freire, P. (1985). *The Politics of Education: Culture, Power and Liberation.* South Hadley, MA: Bergin & Garvey.

Frye, J.R. (2017). Organizational Pressures Driving the Growth of Contingent Faculty, *New Directions for Institutional Research*, Winter: 27–39.

Gappa, J.M., and Leslie, D.W. (1993). *The Invisible Faculty: Improving the Status of Part-Timers in Higher Education.* San Francisco, CA: Jossey-Bass.

Geron, K., and Reevy, G.M. (2018). California Sate University, East Bay: Alignment of Contingent and Tenure Track Faculty Interests and Goals. In Tolley, K. (ed.), *Professors in the Gig Economy: Unionizing Adjunct Faculty in America* (pp. 172–186). Baltimore, MD: Johns Hopkins University Press.

Gilbert, C., and Heller, D. (2010). *The Truman Commission and its Impact on Federal Higher Education Policy from 1947–2010.* Report from Penn State University.

Gilpin, T. (2020). *The Long Deep Grudge.* Chicago, IL: Haymarket Books.

Giroux, H.A. (2008–09). Academic Unfreedom in America: Rethinking the University as a Democratic Public Sphere, *Works and Days*, Vol. 51–54, No. 26/27: 45–71.

Giroux, H.A. (2013). *America's Education Deficit and the War on Youth.* New York: Monthly Review Press.

Giroux, H.A. (2014). *Neoliberalism's War on Higher Education.* Chicago, IL: Haymarket Books.

Goldenberg, E.N., and Cross, J.G. (2011). *Off-track Profs: Nontenured Teachers in Higher Education.* Cambridge, MA: MIT Press.

Goldstene, C. (2012). The Politics of Contingent Academic Labor, *Thought & Action*, Vol. 28 (Fall): 7–15.

Gramsci, A. (1971). *Selections from the Prison Notebooks,* "Wave of Materialism" and "Crisis of Authority." English translation, New York: International Publishers.

Greene, J. (2021). Rethinking the Boundaries of Class: Labor History and Theories of Class and Capitalism, *Labor: Studies in Working Class History*, Vol. 18, No. 2 (May).

Grubb, N.W., and Associates (1999). *Honored but Invisible: An Inside Look at Teaching in Community Colleges.* New York: Routledge.

Haber, G. (2013). *Adjunctivitis.* https://www.gordonhaber.net/adjunctivitis-a-novella/ (available as a Kindle; accessed April 22, 2021).

Hagopian, J. (n.d.). A People's History of the Chicago Teachers Union, *International Socialist Review*, Issue 86. https://isreview.org/issue/86/peoples-history-chicago-teachers-union (accessed April 22, 2021).

Hanson, C., Paavo, A., and Sisters in Labour Education (2019). *Cracking Labour's Glass Ceiling: Transforming Lives through Women's Union Education.* Black Point, Nova Scotia: Fernwood Publishing.

Herbert, W.A., and Apkarian, J. (2017) Everything Passes: Everything Changes: Unionization and Collective Bargaining in Higher Education, *Perspectives on Work*, Labor and Employment Research Association.

Herbert, W.A., and Apkarian, J. (2019). You've Been With the Professors: An Examination of Higher Education Work Stoppage Data, Past and Present, *Employee Rights and Employment Policy Journal*, Vol. 23. Paper preview copy in Joe Berry's collection, 28 pages. Appendix contains table of 43 work stoppages among graduate assistants (7), Tenure Line (1), TT-NTT (8), Non-Academic (21) and NTT (6).

Hess, J. (2003). Using the Internet for Contingent Faculty Organizing, *Jump Cut*, No. 46. www.ejumpcut.org/archive/jc46.2003/links.html (accessed April 22, 2021).

Hess, J., and Hoffman, E. (2014). Organizing for Equality within the Two Tier system. In Hoeller, K. (ed.), *Equality for Contingent Faculty: Overcoming the Two-Tier System.* Nashville, TN: Vanderbilt University Press.

Hess, M. (2003) *Big Wheel at the Cracker Factory* (novel). Morris Publishing. ISBN 10: 1891053078 / ISBN 13: 9781891053078.

Hoeller, K. (2007). The Future of the Contingent Faculty Movement, *Inside Higher Ed*, November: 1–6.

Hoeller, K. (ed.) (2014). *Equality for Contingent Faculty: Overcoming the Two-Tier System.* Nashville, TN: Vanderbilt University Press.

Hoffman, E. (2006). How Mario Savio and Joe Berry Helped CFA Learn to Organize, *California Faculty*, Spring: 19–20.

Hoffman, E., and Hess, J. (2004). Contingent and Faculty Organizing in CFA: 1975–2005. Conference on Contingent Academic Labor VI, Chicago, IL (August).

Hoffman, E., and Hess, J. (2005). Humiliation, Commonality of Interests and the Future of the CSU, *California Faculty*, Fall: 12–14.

Horton, M., with Kohl, J., and Kohl, H. (1990) *The Long Haul, an Autobiography.* New York: Doubleday/Dell.

Hurley, J.H. (2006). *The Adjunct* (novel). Croesus Books.

INCITE! Women of Color Against Violence (2017). *The Revolution Will Not Be Funded: Beyond the Non-Profit Industrial Complex.* Durham, NC: Duke University Press.

James, C.L.R. (1938). *The Black Jacobins: Toussaint L'Ouverture and the San Domingo Revolution*. London: Secker and Warburg.

Jaschik, S. (2015). A Big Union Win, *Inside Higher Education* (January 2). www. insidehighered.com/news/2015/01/02/nlrb-ruling-shifts-legal-ground-faculty-unions-private-colleges (accessed May 5, 2021).

Johnson, B., Kavanagh, P., and Mattson, K. (eds.) (2003). *Steal This University: The Rise of the Corporate University and the Academic Labor Movement*. New York: Routledge.

Jump Cut: A Review of Contemporary Cinema. Looking at media in its social and political context, pioneers since 1974, analyzing media in relation to class, race and gender. www.ejumpcut.org/archive (accessed April 22, 2021).

Karabel, J., and Brint, S. (1989). *The Diverted Dream: Community Colleges and the Promise of Educational Opportunity in America, 1900–1985*. New York: Oxford University Press.

Karpf, J. (2016). Negotiating Over Job Security for Contingent Faculty: The CFA Experience, *Journal of Collective Bargaining in the Academy*, Vol. 11: 13.

Kelley, R.D.G. (1990). *Hammer and Hoe: Alabama Communists during the Great Depression*. Chapel Hill, NC: University of North Carolina Press.

Kerlinger, J. (1998). Success Story: Why Lecturers Are Joining CFA in Record Numbers, *California Faculty*, December: 10.

Kerlinger, J., and Sibary, S. (1998). Protecting Common Interests of Full- and Part-Time Faculty, *Thought And Action, The NEA Higher Education Journal*, Vol. 14., No. 2: 91–102 (special issue: Issues in the Profession: Adjuncts in the Academy. See entire special section of that issue, on adjuncts in the academy: pp. 51–103.)

Kerr, C. (2001). *The Uses of the University*. Cambridge, MA: Harvard University Press.

Kerchner, C.T., Koppich, J.E., and Weeres, J.G. (1998) *Taking Charge of Quality: How Teachers and Unions Can Revitalize Schools, An Introduction and Companion to United Mind Workers*. San Francisco, CA: Jossey-Bass.

Kezar, A., and Maxey, D. (eds.) (2016). *Envisioning the Faculty for the 21st Century: Moving to a Mission-oriented and Learner-centered Model*. New Brunswick, NJ: Rutgers University Press.

Kezar, A., and Sam, C. (2010). Beyond Contracts: Non-Tenure Track Faculty and Campus Governance, *The NEA 2010 Almanac of Higher Education* (pp. 83–92). Washington, DC: National Education Association.

Kociemba, D. (2014). Overcoming the Challenges of Contingent Faculty Organizing, *Academe*, Vol. 100, No. 5: 10–17.

Kolb, D.A. (1984). *Experiential Learning: Experience as the Source of Learning and Development*. Englewood Cliffs, NJ: Prentice-Hall.

Krupat, K., and Tanenbaum, L. (2002). A Network for Campus Democracy: Reflections on NYO and the Academic Labor Movement, *Social Text*, 70, Vol. 20, No. 1 (Spring): 27–50.

Kudera, A. (2010). *Fight for Your Long Day* (novel). Atticus Books. ISBN: 97809 84510504.

Kwok, C.Y. (2018). Psychological Experiences of Contingent Faculty in Oppressive Working Conditions, *Journal of the Professoriate*, Vol. 9, No. 2.

Labaree, D. (1997). Private Goods, Public Goods: The American Struggle Over Education Goals, *American Educational Research Journal*, Vol. 34, No. 1 (Spring): 39–81.

Lave, J., and Wenger, E. (1991). *Situated Learning: Legitimate Peripheral Participation*. Cambridge, UK: Cambridge University Press.

Lavin, D.E., and Hyllegard, D. (1996). *Changing the Odds: Open Admissions and the Life Chances of the Disadvantaged*. New Haven, CT: Yale University Press.

Lau, J.M. (2019). Report on the Coalition Of Contingent Academic Labor (COCAL) XIII Conference, *PS: Political Science & Politics*, Vol. 52, No. 3: 518–519.

Lechuga, M.R., and Ramos, A. (2012). *Magister Changarrization: Los Nuevos sujetos academicos y el trabajo precario en la education superior*. Mexico: STUNAM/Red TAP/Cultura, Trabajo y Democracia/GIIS.

LeFlore, M.A. (2018). Contingent Faculty Unionization: The Impact of Collective Bargaining on Course Pay, Benefits, and Contract Lengths for Part-time Faculty at Four-year Institutions in the US. MA Capstone Paper in Policy Studies, Washington ResearchWorks Archive. http://hdl.handle.net/1773/42753 (accessed April 22, 2021).

Lemann, N. (1999). *The Big Test: The Secret History of the American Meritocracy*. New York: Farrar, Strauss and Giroux.

Leslie, D.W., and Ikenberry, D.J. (1979). Collective Bargaining and Part Time Faculty: Contract Content, *Journal of College and University Personnel Association*, Vol. 30, No. 3 (Fall): 18–26.

Lieberwitz, R.L. (2013). Navigating Troubled Waters at the NLRB, *Academe*, November–December. www.aaup.org/import-tags/yeshiva-ruling (accessed January 2, 2021).

Livingstone, D.W. (2006). Contradictory Class Relations in Work and Learning: Some Resources for Hope (pp. 145–159). In Sawchuck, P.H., Duarte, N., and Elhammoumi, M. (eds.), *Critical Perspectives on Activity: Explorations Across Education, Work & Everyday Life*. Cambridge, UK: Cambridge University Press.

Lustig, J. (2000). Perils of the Knowledge Industry: How a Faculty Union Blocked an Unfriendly Takeover (pp. 319–341). In White, G.D., and Hauck, F.C. (eds.), *Campus, Inc.: Corporate Power in the Ivory Tower*. Amherst, NY: Prometheus Books.

Lyons, R.E., Kysilka, M.L., and Pawlas, G.E. (1999). *The Adjunct Professor's Guide to Success: Surviving and Thriving in the College Classroom*. Boston, MA: Allyn and Bacon.

MacLean, N. (2017). *Democracy in Chains: The Deep History of the Radical Right's Stealth Plan for America*. New York: Viking.

Magolda, P.N. (2016). *The Lives of Campus Custodians: Insights into Corporatization and Civic Disengagement in the Academy*. Sterling, VA: Stylus.

Maitland, C. (1985). The Campaign to Win Bargaining Rights for the California State University Faculty. Unpublished dissertation, Claremont Graduate University, CA.

Maitland, C., and Rhoades, G. (2005). Bargaining for Contingent Faculty, *The NEA 2005 Almanac of Higher Education*. Washington, DC: National Education Association.

Martin, D. (1995). *Thinking Union: Activism and Education in Canada's Labour Movement*. Toronto: Between the Lines.

Martin, R. (ed.) (1998). *Chalk Lines: The Politics of Work in the Managed University*. Durham, NC: Duke University Press.

Mazurek, R.A. (2011). Academic Labor is a Class Issue: Professional Organizations Confront the Exploitation of Contingent Faculty, *Journal of Workplace Rights*, Vol. 16, No. 3–4: 353–366.

McAlevey, J.F. (2012) *Raising Expectations and Raising Hell: My Decade Fighting for the Labor Movement*. London: Verso.

McAlevy, J.F. (2016) *No Shortcuts: Organizing for Power in the New Gilded Age*. New York: Oxford University Press.

McGee, M. (2002). Hooked on Higher Education and Other Tales from Adjunct Faculty Organizing, *Social Text*, 70, Vol. 20, No. 1 (Spring): 61–79.

Metchick, R.H., and Singh, P. (2004). *Yeshiva* and Faculty Unionization in Higher Education, *Labor Studies Journal*, Vol. 28, No. 4: 45–65.

Mitchell, D.E., and Nielsen, S.Y. (2012). Internationalization and Globalization in Higher Education (pp. 3–22). In Cuadra-Montiel, H. (ed.), *Globalization – Education* and *Management Agendas*. Rijeka, Croatia: InTech.

Moran, M. (September 2016). MyVU. https://news.vanderbilt.edu/2016/09/19/university-budgeting-system-evolves-to-support-strategic-priorities/ (accessed January 2, 2021).

Moser, R. (2000). The AAUP Organizes Part-Time Faculty, *Academe*, Vol. 86, No. 6 (November–December): 34–37.

Moser, R. (2004) Campus Democracy, Community, and Academic Citizenship. Paper presented at the Conference on Contingent Academic Labor VI, Chicago, IL, August 2004.

Moser, R. (2020). The Inside/Outside Strategy Revisited, *BeFreedom: Movement Strategy* (March 6). https://befreedom.co/inside-outside-strategy-revisited/ (accessed December 20, 2020).

Munene, I.I. (2018). *Contextualizing and Organizing Contingent Faculty*. Lanham, MD: Lexington Books.

Murphy, M. (1990). *Blackboard Unions: The AFT and the NEA, 1900–1980*. Ithaca, NY: Cornell University Press.

Myerhoff, E., and Noterman, E. (2019). Revolutionary Scholarship by Any Speed Necessary: Slow or Fast but for the End of This World, *ACME, An International Journal for Critical Geographies,* Vol. 18, No. 1: 217–245.

National Center for the Study of Collective Bargaining in Higher Education. (January 2006). Moriarty, J., Savarese, M., and Boris, R. (eds.). *Directory of US Faculty Contracts and Bargaining Agents in Institutions of Higher Education.* Series II, No. 1.

National Center for the Study of Collective Bargaining in Higher Education. (September 2012). Savarese, M., and Berry, J. (eds.) *Directory of U.S. Faculty Contracts and Bargaining Agents in Institutions of Higher Education.* Series II, No. 2.

National Center for the Study of Collective Bargaining in Higher Education. (2020). Herbert, W.A., Apkanian, J., and van der Haald, J. (eds.) *Supplementary Directory of New Bargaining Agents and Contract in Institutions of Higher Education.*

Nelson, C. (ed.) (1997). *Will Teach for Food: Academic Labor in Crisis.* Minneapolis, MN: University of Minnesota Press.

Nelson, C., and Watt, S. (1999). *Academic Keywords: A Devil's Dictionary for Higher Education.* New York: Routledge.

Newfield, M. (2006). Contingent Faculty Bargaining: Separate but Equal? *Journal of Collective Bargaining in the Academy,* No. 1: 15.

Newfield, M., Berry, J., and Kroik, P. (eds.) (2015). Contingent Academic Labor: The Way Forward, *WorkingUSA: The Journal of Labor and Society,* Vol. 18, No. 3 (September) (special issue on contingent faculty).

Newman, M. (1993). *The Third Contract: Theory and Practice in Trade Union Training.* Paddington, New South Wales: Stewart Victor Publishing.

Newman, M. (1994). *Defining the Enemy: Adult Education in Social Action.* Paddington, New South Wales: Stewart Victor Publishing.

Newman, M. (1999). *Maeler's Regard: Images of Adult Learning.* Paddington, New South Wales: Stewart Victor Publishing.

Newman, M. (2006). *Teaching Defiance: Stories and Strategies for Activist Educators, a Book Written in Wartime.* San Francisco, CA: Jossey-Bass/Wiley.

Norton, T.M., and Ollman, B. (1978). *Studies in Socialist Pedagogy.* New York: Monthly Review Press.

Ovetz, R. (2015). Migrant Mindworkers and the New Division of Academic Labor, *WorkingUSA,* Vol. 18, No. 3: 331–347.

Palmquist, M., and Doe, S. (2011). Contingent Faculty: Introduction, *College English,* Vol. 73, No. 4: 353–355. www.jstor.org/stable/23052344 (accessed February 25, 2021).

Parker, M., and Gruelle, M. (1999). *Democracy is Power: Rebuilding Unions from the Bottom Up.* Detroit, MI: Labor Notes.

Poirier, T. (2018). *Non-Regular.* Vancouver, BC: U/P UNIT/PITT Projects.

Powell, L.F., Jr. (1971). *Memo to Chairman, Education Committee.* US Chamber of Commerce.

Pressgang Publishers, Toronto (1978). Class Consciousness, Class Analysis (poster). http://collections.museumca.org/?q=collection-item/20105420125 (accessed April 22, 2021).

Public Policy Institute of California: Proposition 13. www.ppic.org/blog/tag/proposition-13/ (accessed December 30, 2020).

Rainey, R., and Kullgren, I. (2019), 1 year after Janus, unions are flush, *Politico* (May 17), www.politico.com/story/2019/05/17/janus-unions-employment-1447266 (accessed May 7, 2021).

Readings, B. (1996). *The University in Ruins.* Cambridge, MA: Harvard University Press.

Reichman, H. (2019). Do Adjuncts Have Academic Freedom? or Why Tenure Matters: The Costs of Contingency (October). www.aaup.org/article/do-adjuncts-have-academic-freedom-or-why-tenure-matters? (accessed on January 2, 2020).

Rein, M., Ellinger, M., and Legion, V. (2021) *Free City! The Fight for San Francisco City College and Education for All.* Oakland, CA: PM Press.

Rhoades, G. (1998). *Managed Professionals: Unionized Faculty and Restructuring Academic Labor.* Albany, NY: State University of New York Press.

Rhoades, G. (2009). Carnegie, DuPont Circle and the AAUP: (Re)Shaping a Cosmopolitan Locally-Engaged Professoriate. *Change Magazine*, January–February (author's print, no page available).

Rhoades, G. (2013). Disruptive Innovations for Adjunct Faculty: Common Sense for the Common Good, *Thought & Action*, Vol. 29 (Fall): 71–86.

Rhoades, G. (2015). Creative Leveraging in Contingent Faculty Organizing, *WorkingUSA*, Vol. 18, No. 3: 435–445.

Rhoades, G. (2019). Taking College Teachers' Working Conditions Seriously: Adjunct Faculty and Negotiating a Labor-based Conception of Quality, *Journal of Higher Education*, Vol. 91, No. 4: 1–26.

Rhoades, G., and Slaughter, S. (1997). Academic Capitalism, Managed Professionals, and Supply-side Higher Education, *Social Text*, Vol. 51: 9–38.

Rose, M. (2004) *The Mind at Work: Valuing the Intelligence of the American Worker.* New York: Viking.

Rousmaniere, K. (2005). *Margaret Haley, Citizen Teacher: The Life and Leadership of Margaret Haley.* New York: SUNY Press.

Schell, E.E. (1998). *Gypsy Academics and Mother-Teachers: Gender, Contingent Labor, and Writing Instruction.* Portsmouth, NH: Boynton-Cook.

Schell, E.E., and Stock, P.L. (eds.) (2001). *Moving a Mountain: Transforming the Role of Contingent Faculty in Composition Studies and Higher Education.* Urbana, IL: National Council of Teachers of English.

Schell, E.E. (2001). Toward a New Labor Movement in Higher Education: Contingent Labor and Organizing for Change, *Workplace: A Journal for Academic Labor*, Vol. 4, No. 1.

Schneirov, R. (2003). Contingent Faculty: A New Social Movement Takes Shape, *WorkingUSA*, Vol. 6, No. 4: 38–48.

Schrecker, E.W. (1986). *No Ivory Tower: McCarthyism and the Universities.* New York: Oxford University Press.

Schwartz, J.M. (2014). Resisting the Exploitation of Contingent Faculty in the Neo-liberal University: The Challenge of Building Solidarity Between Tenured and Non-tenured Faculty, *New Political Science*, Vol. 36, No. 4 (October 30): 504–522. http://dx.doi.org/10.1080/07393148.2014.9548 (accessed April 22, 2021).

Schwartz, R.M. (1999, 2003). *The Labor Law Source Book: Texts of Federal Labor Laws.* Cambridge, MA: Workrights Press.

Semuels, A. (2018). Is This the End of Public Sector Unions in America? *The Atlantic* (June 27). https://www.theatlantic.com/politics/archive/2018/06/janus-afscme-public-sector-unions/563879/ (accessed May 3, 2021).

Schwartz, R.M. (2012). *No Contract, No Peace!* Published in 2006 as *Strikes, Picketing and Inside Campaigns.* Boston, MA: Workrights Press.

Shor, I. (1996) *When Students Have Power: Negotiating Authority in a Critical Pedagogy.* Chicago, IL: University of Chicago Press.

Shulman, S. (2014) *Faculty and Graduate Student Employment Report, 2014* (April). Department of Economics, Center for the Study of Academic Labor, Colorado State University, Fort Collins, CO 80523.

Slaughter, S., and Leslie, L.L. (1997). *Academic Capitalism: Politics, Policies, and the Entrepreneurial University.* Baltimore, MD: Johns Hopkins University Press.

Smith, A. (1776 [1998]). *An Inquiry into the Nature and Causes of the Wealth of Nations, a Selected Edition,* K. Sutherland (ed.). Oxford: Oxford University Press.

Smith, S.A. (2015). Contingent Faculty and Academic Freedom in the Twenty-First Century, *First Amendment Studies*, Vol. 49, No. 1: 27–30. DOI: 10.1080/21689725.2015.1016362 (accessed April 22, 2021).

Soley, L.C. (1995). *Leasing the Ivory Tower: The Corporate Takeover of Academia.* Boston, MA: South End Press.

Sultanova, A., and Ivanova, I. (2011). The Features of a Social Situation of Development of Children under Modern Russian Conditions, *Tätigkeitstheorie: E-Journal for Activity Theoretical Research in Germany*, No. 4: 53–64.

Tam, T., and Jacoby, D. (2009). What We Can't Say About Contingent Faculty, *Academe*, Vol. 95, No. 3: 19–22.

Task Force on Temporary Faculty (1977). *Report of the Task Force on Temporary Faculty.* Office of the Chancellor, California State University System.

Thompson, K. (2003). Contingent Faculty and Student Learning: Welcome to the Strativersity, *New Directions for Higher Education*, Vol. 2003, No. 123: 41–47.

Tirelli, V. (1997). Adjuncts and More Adjuncts: Labor Segmentation and the Transformation of Higher Education, *Social Text*, Vol. 51 (Summer): 75–91.

Tirelli, V. (2007). The Invisible Faculty Fight Back: Contingent Academic Labor and the Political Economy of the Corporate University. Unpublished dissertation submitted to the Graduate Faculty in Political Science in partial fulfillment of the requirements for the degree of Doctor of Philosophy, The City University of New York, NY.

Tirelli, V. (2014). Contingent Academic Labor Against Neoliberalism, *New Political Science*, Vol. 36, No. 4: 523–537.

Tokarczyk, M.M., and Fay, E.A. (eds.) (1993). *Working Class Women in the Academy: Laborers in the Knowledge Factory.* Amherst, MA: University of Massachusetts Press.

Tolley, K. (ed.) (2018). *Professors in the Gig Economy: Unionizing Adjunct Faculty in America.* Baltimore, MD: Johns Hopkins University Press.

Truman Commission on Higher Education (1947). *Higher Education for Democracy: A Report of the President's Commission on Higher Education*, Vol. 1, *Establishing the Goals.* New York.

Turk, J.L. (ed.) (2008). *Universities at Risk: How Politics, Special Interests, and Corporatization Threaten Academic Integrity.* Toronto, Canada: James Lorimer.

UC Berkeley Library (2000). The Heart of the 1960 California Masterplan for Higher Education. Adapted from Douglass, J., *The California Idea and American Higher Education, 1860 to the 1960 Master Plan* (Stanford, CA: Stanford University Press, 2000). www.lib.berkeley.edu/uchistory//archives_exhibits/masterplan/heartmp.html (accessed January 2, 2020).

Unger, D.N.S. (2000). Academic Apartheid: The Predicament of Part time Faculty (from Spring 1995), *Thought and Action: The NEA Higher Education Journal*, Vol. 16, No. 2 (Fall): 61–64.

US Government (1944). The GI Bill of Rights: An Analysis of the Serviceman's Readjustment Act of 1944, *Social Security Bulletin* (July). www.ssa.gov/policy/docs/ssb/v7n7/v7n7p3.pdf (accessed December 30, 2020).

Wagoner, J., and Snow, D. (2011). *The Freeway Flier and the Life of the Mind* (novel). XLibris Corporation at XLibris.com. ISBN-10: 1456831194 / ISBN-13: 978-1456831196.

Wallace, M.E. (ed.) (1984). *Part-Time Academic Employment in the Humanities.* New York: Modern Language Association.

Walton, R.E., and McKersie, R.B. (1965). *A Behavioral Theory of Labor Negotiations.* New York: McGraw-Hill.

Weingarten, K. (2004). *Common Shock: Witnessing Violence Every Day.* New York: New American Library.

Wells, J. (1991). *The Part-Time Teacher* (poetry). Willits, CA: Rainy Day Women Press.

Wiegard, A. (2015). The Birth of a Contingent Faculty Activist: Interview with Kim Laffont, *WorkingUSA*, Vol. 18, No. 3: 487–495. (See also this entire special issue of *WorkingUSA* devoted to contingent faculty.)

Willis, E. (2002). The Post-*Yeshiva* Paradox: Faculty Organizing at NYU, *Social Text*, Vol. 20, No 1 (Spring): 11–25.

Wilson, J.K. (2008). *Patriotic Correctness: Academic Freedom and Its Enemies.* Boulder, CO: Paradigm Publishers.

Wootton, L., and Moomau, G. (2017). Building Our Own Bridges: A Case Study in Contingent Faculty Self-Advocacy (pp. 199–212). In Kahn, S., Lalicker, W., and Lynch-Biniek, A. (eds.), *Contingency, Exploitation, and Solidarity: Labor and Action in English Composition.* Boulder, CO: WAC Clearinghouse.

Worthen, H. (2001). The Problem of the Majority Contingent Faculty in the Community Colleges (pp. 42–61). In Alford, B., and Kroll, K. (eds.), *The Politics of Writing in the Two Year College*. Portsmouth, NH: Boynton-Cook.

Worthen, H. (2014). *The Status of Labor Education in Higher Education in the U.S.* https://uale.org/resource/links/ (accessed April 22, 2021).

Worthen, H. (2014). *What Did You Learn at Work Today? Forbidden Lessons of Labor Education*. Brooklyn, NY: Hardball.

Worthen, H. (2015). Organizing as Learning: The Metro Strategy and a Community of Practice for Faculty, *WorkingUSA*, Vol. 18, No. 3: 421–434.

Worthen, H., and Berry, J. (1999). *Contingent Faculty in Public Higher Education in Pennsylvania, Spring 1999: Focus on Community Colleges*. A report presented at the annual meeting of the Faculty Coalition on Higher Education, April 30–May 1, Harrisburg, PA. Keystone Research Center, 412 North Third Street, Harrisburg, PA 19101.

Worthen, H., and Berry, J. (2006). Our Working Conditions are Our Students' Learning Conditions: A CHAT Analysis of College Teaching (pp. 145–159). In Sawchuk, P., Duarte, N., and El Hammoumi, M. (eds.), *Critical Perspectives on Activity: Explorations Across Education, Work & Everyday Life*. Cambridge, UK: Cambridge University Press.

Wright, E.O. (1976). Class Boundaries in Advanced Capitalist Societies, *New Left Review*, No. 98 (July–August): 3–41.

Wright, E.O. (1978). *Class, Crisis and the State*. New York: Schocken Books.

Young, M.D. (1968). *The Rise of the Meritocracy, 1870–2033, an Essay on Teaching and Equality*. Harmondsworth, UK: Pelican.

Zabel, G. (April, 2000). A New Campus Rebellion: Organizing Boston's Contingent Faculty, *New Labor Forum* (pp. 90–98). Labor Resource Center, Queens College, City University of New York.

Zabel, G. (2016–17). Critical Revolutionary Praxis in the Neoliberal University. In double issue of *Works and Days* and *Cultural Logic*, pp. 187–209. https://ices.library.ubc.ca/index.php/workplace/article/view/186386 (accessed April 22, 2021).

Zabel, G., and Barbara, G. (2003). Social Movement Unionism and Adjunct Faculty Organizing in Boston. In Mattson, K., Heber Johnson, B., and Kavanagh, P. (eds.), *Steal This University: The Rise of the Corporate University and the Academic Labor Movement*. New York: Routledge.

Zheng, R. (2018). Precarity is a Feminist Issue: Gender and Contingent Labor in the Academy, *Hypatia*, Vol. 33, No. 2: 235–255.

Zinn, H. (1980). *A People's History of the United States: 1492–Present*. New York: Harper and Row.

Zook, G.P. (1947), *Higher Education for American Democracy: A Report on the President's Commission on Higher Education*, Vols I–VI. "The Truman Report." Washington, DC: US Department of Health, Education and Welfare, US Government.

Index

note: *ill* refers to an illustration; *n* to a note; *t* to a table

faculty of color 179
faculty unions 90, 228–9
Fair Share Act (2000) 49–51, 53, 55, 204
  see also agency fee/agency shop
Fendel, Nina 22, 27, 34, 43, 53, 59
  and CFA election 39
  and CFA planning meeting 56
  and Fair Share election 53
  and layoffs 28–9, 30, 122
"first consideration": and hiring policy
  102
free speech 105, 235
Free Speech Movement (FSM) 212
Freedom Summer 75, 140, 255*n*6
Freeway Flyer 215
French Revolution 166
Future of the University Conference
  58, 59

gay rights movement 194, 260–1*n*2
General Motors strike 210
general strikes (1934) 210
George Washington University 147
GI Bill 4, 71–2, 74
gig economy 243
Giles, Douglas 106–7
Global Financial Crisis (2008) 30
Gluck, Sherna 26
Goldwater, Barry 141
governance 59, 60, 129, 243–4
  AAUP Statement on 186
  shared governance 100, 182, 186, 187,
    201, 243–4, 248
graduate employees 176, 205
graduate schools 84–5
Gramsci, Antonio 233
  Prison Notebooks 133
Green, William 217
grievances: resolution of 173–4
Gulf War (1991) 30

Haitian Revolution 166–7, 195
Hamer, Fannie Lou 141, 142

Hanson, Cindy and Adriane Paavo
  Cracking Labour's Glass Ceiling 218
Hayakawa, S.I. 14
Healy, Margaret 183
Hegel, Georg Wilhelm Friedrich 155–6
hegemony 133, 244
  cultural hegemony 133, 170, 182
Henderson, Billy 171
Herbert, William A. and Jacob
  Apkarian 176
Hess, John 6–7, 11, 31–2, 33, 58, 187,
  191, 211, 215, 249–50
  and I/O strategy 136
  asks Stuart Long to leave Lecturers'
    meeting 227
  and role-play training 192
  at San Francisco State 12–13, 52–3,
    164
  on tenure-line faculty 65
higher education 3, 13, 69–70, 157
  expansion of 71–4
  Humboldt model of 71
  transitions in 4–5, 69–70, 86
  see also universities
Higher Education Employee-Employer
  Relations Act (HEERA) (1979) 18,
  49, 114
Hoeller, Keith *Equality for Contingent
  Faculty* 212
Hoffa, Jimmy Jr 33, 53
Hoffman, Elizabeth 20, 25–7, 40, 41–2,
  45–6, 49, 164, 194
  and emeritus status 104–5
  on Lecturers Council 54
  on merit pay 93
  and role-play training 192
Horton, Myles 163
How to Jumpstart Your Union 211
  see also Labor Notes
Humphrey, Hubert 141–2

imposter syndrome 190
Industrial Workers of the World (IWW)
  16, 167, 251*n*4

Thanks to our Patreon Subscribers:

Lia Lilith de Oliveira
Andrew Perry

Who have shown generosity and
comradeship in support of our publishing.

Check out the other perks you get by subscribing
to our Patreon – visit patreon.com/plutopress.

Subscriptions start from £3 a month.

The Pluto Press Newsletter

Hello friend of Pluto!

Want to stay on top of the best radical books
we publish?

Then sign up to be the first to hear about our
new books, as well as special events,
podcasts and videos.

You'll also get 50% off your first order with us
when you sign up.

Come and join us!

Go to bit.ly/PlutoNewsletter